REVISED & **2nd EDITION** EXPANDED

ELECTRONIC DESIGN AND PUBLISHING

BUSINESS PRACTICES

BY
LIANE
SEBASTIAN

NEW YORK CITY

ALLWORTH
PRESS

▷ This book is dedicated to all publishing pioneers on the bleeding edge.

This book is especially dedicated to the memory of Michael Waitsman who passed away as it was at the printer. As one of Michael's last design projects, he was the spirit behind this effort. He brought the Chicago publishing community together in 1986 to pioneer desktop publishing. As a visionary, designer, photographer, writer, inventor, and one of the country's first multimedia artists, Michael was a great influence on our profession. He inspired and mentored many professionals. It was Michael who saw the need for this book and helped me with the project. He died on February 19, 1995 at age 47 of a brain tumor. Some say the good die young. Michael's talents and integrity made him one of our industry's best. His reach will be felt forever. To my partner and husband, I dedicate this second edition, for without him, I could never have created this work.—Liane Sebastian, March 19, 1995.

Published by Allworth Press, an imprint of Allworth Communications, Inc., 10 East 23rd Street, New York, NY 10010. Distributor to the trade in the United States: Consortium Book Sales & Distribution, Inc., 287 East Sixth Street, Suite 365, Saint Paul, MN 55101.

Library of Congress Catalog Card Number: 92–71564
ISBN: 1–880559–03–X

Book design and cover production by Michael Waitsman.
Interior production by Laurel Dantzig, and Rose Sweet.
Synthesis Publishing
7352 North Winchester
Chicago, Illinois, 60626, 312/609-1111

ACKNOWLEDGMENTS

EDITORS
Betsy Shepherd, freelance writer, Chicago
Michael Waitsman, Synthesis Concepts Inc.
Van Tanner, Association of Imaging Service Bureaus
Eda Warren, Desktop Publishing Services.

REVIEWERS
Ardath Berliant, Freelance Hotline, Inc./Desktop Temps, Inc.
Gregg Cygan, H+W Graphics
Tad Crawford, Allworth Press
Barbara Hall, College of DuPage
Susan Wascher-Kumar, Information Design Services.

COUNSEL
William T. McGrath, Davis, Mannix & McGrath.

SUPPORTERS
ADEPT (Association for the Development of Electronic Publishing Technique)
AIGA (American Institute of Graphic Arts—Chicago)
AISB (Association of Imaging Service Bureaus)
APA (Advertising Photographers of America)
IABC Chicago (Int'l Association of Business Communicators)
IWOC (Independent Writers of Chicago).

SOURCES
GCA (Graphic Communications Association); GAG (Graphic Artists Guild); PIA (Printing Industries of America).

SPECIAL THANKS
Clay Bodine, Desktop Resource Inc.; Collin Canright, Canright Writing; Barrie Dellenbach, Miller, Mason & Dickenson; Eric Diamond, EPR; Matthew Doherty, Matthew Doherty Design; David Doty, PageWorks; Barbara Golden, Computing Solutions; Jim Maivald, Desktop Design; Earl McGhee, McGhee Consulting Group; Tom Olson; Jim Madden, Rider Dickerson, Inc.; John Snyder, Snyder Photography; Jeff Schewe, Schewe Photography. Finally, I would like to especially thank Laurel Dantzig, Barbara Bennecke, Marty Turck, and Michael Waitsman, all of Synthesis Concepts, Inc., for putting up with me and the work.

CONTENTS

PERSPECTIVE

PLANNING

IMAGES

DEVELOPMENT

PRODUCTION

COMPLETION

OWNERSHIP

APPENDIX

PERSPECTIVE

PREFACE

Electronic Design and Publishing: Business Practices is a guide to procedure in the quickly changing publishing industry. It is not law, though several sections are based on laws that apply to our industry. It is not a book of musts but rather of suggestions to help you avoid the most common problems that arise in a publishing project.

Communication is a challenge in any industry, but more so in the communications field. Desktop publishers are continually moving into unknown territory. The purpose of this book is to help you find the best practices for your own working methods. Adapt it, use it as a springboard for thought, agree with it, disagree with it, challenge it—but most importantly, let it inspire and improve your procedures.

Although this book is written by one person, it has been reviewed by professionals throughout the industry. The Board of Directors of the Association for the Development of Electronic Publishing Technique (ADEPT), who represent ADEPT's membership, endorsed the origins of this text, but recognizes the great variety and breadth of the publishing industry. If you have improvements you would like to contribute to this effort, we would like to hear about them.

We hope that all in the industry find these *Publishing Practices* to be instrumental in planning their publishing projects—again and again.

FOREWORD

It's hard to believe it has only been eight years since I first worked with a Macintosh. (The significance of the Mac is that without it there would be no desktop publishing). As President of AISB (Association of Imaging Service Bureaus), I have witnessed firsthand the changes that have occurred in the design and publishing industry. Typesetters, color separators, and service bureaus are all merging into one business: "Imaging centers" (a term AISB uses to refer to the companies that have successfully brought together the new high-tech imaging with the traditional crafts of type and color). These are providing a smorgasbord of services designed to meet changing demands.

Crafts that were kept distinctly separate are now overlapping. This is causing confusion, not only for the people being reskilled, but for the customers of these services. This is compounded by the trend of the customer absorbing more prepress tasks in-house. What results is that responsibilities are being shifted or diffused. The new imaging centers are profoundly aware of the need for trade practices. We are the middle persons in the publishing process: We work between the document originators and the printers. If roles are not adequately defined, our job becomes very difficult—if not impossible.

I was delighted Liane sought my guidance in making sure the imaging centers' perspective is represented in this book. The extensive number of drafts and rewrites I participated in assure me that Liane is thorough and fair in her quest to present all sides of the issues. It's obvious that she has painstakingly researched this book, and her efforts have certainly paid off. This book stand as a solid frame of reference for all members of the electronic design community.

Desktop publishing, using a Macintosh, was introduced in the mid-1980s. Most design and prepress professionals paid little attention to its capabilities. It was considered to be a tool for the home and office crowd. Only the true visionaries saw the Mac as more than just a toy, and had a battle trying to convince others to give "desktop" a try. They brown-bagged their Macs into the back doors of the workplace, until finally desktop publishing became respectable. They persevered and caused a whole industry to turn inside-out in less time than it takes to get a college degree.

It's humorous to think back about those who stood up and declared "desktop publishing" to be only for novices. They predicted desktop's demise. Today, these same people are not only acknowledging desktop publishing, they claim to be experts! I believed them to be wrong then and I believe them to be wrong now. They are not the practitioners—and can't see the industry from the trenches. The future still belongs to the visionaries who are making the new system work.

The AISB office receives numerous calls from people in dispute over design projects. Normally it's an imaging center or their client wishing to find out who is to blame when a project has problems. Most of these problems could have been eliminated if the persons involved had followed the Business Practices outlined in this book. However, communication is still paramount. If an imaging center adheres to these Practices, but the clients or designer do not, problems will still ensue. That's why I recommend that everyone in the design and prepress industry read this book. If you're in charge of a project, read this book first, then make sure your clients and suppliers read it too. That way, everyone will march to the same tune.

Van Tanner
President of AISB and
VerTec Solutions, Inc.
aging Center Consulting Firm)
Greensboro, North Carolina)

CLEAR ROLES FOR NAVIGATION

Print communication goals remain consistent through time. However, the way publications such as books, magazines, newsletters, catalogs, brochures, booklets, stationery, posters, flyers, menus, advertisements, and other printed items are created continues to evolve. Beginning with movable type, advancing with offset lithography, and now with computers, publication technology takes another great leap in capabilities. In the computer age, people must change to keep pace with the technology. Managing and adapting to change is critical to success.

The roles of professionals involved in the process are transforming along with the process itself. Although the end-product is substantially the same, responsibilities, roles, and ownerships are being redefined. As the chain becomes increasingly interlinked, questions on these topics arise more often. Perhaps the biggest changes are that the creator of a publication is empowered to do more and the process happens faster.

In the past, responsibilities were clearly defined by sequential phases as a project went from the hands of one expert to another. As phases are compressed, changed, or eliminated, professional roles overlap or are reassigned. As roles and activities become more complex, responsibilities become more confusing.

It helpful to have a comprehensive document the entire industry can rely on to clarify responsibilities. Ideally, this document will cover each step of the process, from the Originating Client's need for a publication, through the many creative, production, financial, and printing stages.

This document is designed to cover all print-oriented media produced through desktop publishing. If it is to be as useful as possible, it may be referred to in legal proceedings. Therefore, two of the reviewers are attorneys with expertise in publishing. Nevertheless, we've put a lot of effort into writing this in plain English rather than legalese. Caution needs to be taken in utilizing this document legally. For it to have any influence in a legal dispute, there must be proven prior knowledge of these practices by all parties involved. References should be made in the reader's own professional agreements that refer to the *Electronic Design and Publishing: Business Practices* to ensure that everyone is aware of responsibilities. Copies of this book can be shared by all those participating in a project at the beginning to facilitate communication.

Forefronts of electonic publishing:
• Hardware—workstations grow in power, notebook computers expand portability, and advances in input methods, such as pen-based, voice activation, and handwriting recognition.
• Workgroup publishing—software and hardware advances in networking capabilities.
• New media publishing—multimedia, integrated publishing, and desktop presentations. Hypertext also offers new ways to interact.
• Font developments—such as Multiple Masters, where the user can customize a typeface.
• Database publishing—personalization of printed pieces through linking databases to page composition, allowing for targeted time-sensitive marketing.
• Document interplatform exchange—such as Editable PostScript or other methods for opening documents created in one environment into another.
• Prepress and color capabilities—including color calibration, compression (color files require vast storage), and color registration (further developments in trapping).
• Direct-to-press printing—going from electronic files straight to printed pieces, which will make printing faster and less expensive.
• New forms of printing—such as stochastic and high fidelity.

Many organizations have contributed previously written specialty trade practices for inclusion into this comprehensive guide. These practices or organizations are listed in the Acknowledgments. We thank them for their support and for helping to make this a thorough document—the first set of guidelines to cover the new processes comprehensively.

While this book deals in part with various legal subjects and contains general information about various aspects of the law as it applies to electronic design and publishing, the author is not a lawyer and this book is not intended to be taken as legal advice. Resolution of legal issues for specific situations depends on the evaluation of precise factual circumstances. For this you should always consult with your attorney.

Many new participants to the publishing process are unaware of traditional trade practices. It has become even more important to define responsibilities in the computerized environment due to shorter schedules, quickly changing technology, and evolving skills. In fact, this lack of knowledge might make professionals unaware of their own responsibilities in a publishing project—and could even provide a loophole should a dispute arise. Therefore, this document functions best when reviewed at the critical phases of each project.

It's far better for there to be rapport and cooperation between all parties throughout a project than to have misunderstandings as the project progresses. This guide can help in building such rapport.

THE INDUSTRY'S MATURING PROCESS

SECOND EDITION

The honeymoon is over and a deeper fascination has set in. The marriage between computers and publishing has evolved into the everyday challenge of learning how to live together. The fascination, once magical and all-consuming, has matured to a symbiotic functionalism. The challenges, however, never become less. There's no room for complacency or getting too comfortable if someone wants to have a viable business that can compete in a shifting and quickly moving climate.

Health requires balance: between the initial romantic fascination of a new technology, to the patience, understanding, and nurturing of a new culture.

Entering electronic publishing's second decade means it's time to collect the lessons earned from experience to take an active role in both ours and the industry's potential. *Electronic Design and Publishing: Business Practices* was begun during the turbulent late eighties and published in 1992, just as the dust was shaking off our boots from a hike through unknown terrain. The path we traveled is more known now. The trends, momentum, new forms of challenge, and a new visual language, have all settled into normalcy. Technical decisions become commonplace.

The more difficult challenges are people—defining roles, turf, control, skills, and responsibilities. The human needs become more obvious.

Talents for maturing and handling the intangibilities of workflow include:

- insight
- appropriateness
- charm and tact
- process experimentation
- perceptiveness
- flexibility
- listening
- management

Talents for maturing and handling the tangibility of technological concerns:

- patience
- scientific focus
- analytical skills
- constant study
- technique experimentation
- interface experimentation
- teaching
- craft

Workflow is less tangible and harder to measure than the tangibility of technology.

Success balances the two individual visions of professionals oriented more to workflow and those oriented more to technological advancements. Both groups also share talents in common as more professionals become skilled in the technology. Needless to say, the cutting edge always has its virtuosos, but the use of computer tools themselves is now pervasive.

The first edition of *Electronic Design and Publishing: Business Practices* was formulated with the help of ADEPT (the Association for the Development of Electronic Publishing Technique)—one of the first user groups in the country. More industry associations are contributing now to its form and content.

This edition has five new chapters, keeping pace with more recently discovered workflow concerns. The quotes that are added from industry voices primarily come from conferences, secondarily from key publications, and then from direct interviews. They have been collected over the last ten years, but all have a ring of timeless wisdom.

Electronic Design and Publishing: Business Practices's future editions will continue to mirror the industry changes, define what remains the same through time, and have an outreach to every facet of electronic publishing.

Talents required to be successful in blending the human concerns with the technological:

- process definition
- questioning
- develop fascination for and love of change
- capabilities expansion
- continuous financing
- planning and projection
- looking at problems in new ways
- understanding of business practices

PLANNING

1 DEFINING ROLES

Trying to categorize the various players on the field of publishing is like trying to guess who will catch a ball before it is thrown. Roles are unclear because traditional definitions are transformed by electronic definitions.

All publishing activities, independent of size, fall into three major functions, groups, or roles: the Client role, the Creative role, and the Print role. Some functions overlap (such as an account manager working with the Creative Group, or a Page Composer working for the Printer), and some roles involve more than one person. Also, one person can perform more than one role (such as a Designer who handles color separations). Sometimes only two people are involved in the entire process (as in is a small project).

Electronic publishing reduces the number of disciplines, and in some ways simplifies the process as each individual "wears more hats," though the growing skills of this individual are quite complex. Electronic publishing also compresses project phases and schedules. Although this is attractive to those needing printed materials quickly, such compression can allow more errors, for there are fewer checkpoints and fewer eyes reviewing each stage. Mistakes cost extra money, so an efficient approval process is the best form of protection against oversights.

One of ADEPT's founding members, Matthew Doherty, advises that the publishing process proceeds most smoothly when those working together keep three guiding "C's in mind:
- Communication
- Common sense
- Courtesy between professionals

The two crucial role links are the intersections between the Originating Client and the Creative Group, and between the Creative Group and the Print Group.

ORIGINATING CLIENT GROUP

The Originating Client Group is the organization or person needing a publication. This group begins the process, ultimately pays the bills, and receives the intended benefit of the material.

Client: The decision-maker (and invoice-approver). This could be a business owner, a marketing and sales professional, or a publisher. Sometimes this role is served by a small committee, such as company officers or directors.

Project Manager (inside the Originating Client organization): The person who interacts with the Creative Group most frequently. This person provides information, sets up meetings, checks quality, and facilitates processes.

CREATIVE GROUP

The Creative Group is an in-house department or independent firm.

Art Director (sometimes the Creative Director and sometimes the Designer): Manages the project and is responsible for the quality of the entire design and production process.

Researcher: Finds all relevant facts, background information, handles any market research, and accesses resources.

Designer: Originates concepts and ideas, facilitates the production process, and coordinates resources.

Writer: Generates the text for the written part of the publication. Usually begins by setting up an outline and reviewing it with the Client, the Art Director, and the Designer.

Editor: Evaluates writing for content and style, makes sure the correct version is assembled for design, checks proofreading, adapts text to the design if necessary, and checks over proofs for typographical errors.

Proofreader: Checks through the manuscript (sometimes called copy) after editing the drafts for typographical and grammatical errors (sometimes a Copy Editor is utilized for consistency of style and word usage).

Illustrator or Photographer: Generates original visual images for use in a publication, in accordance with the design.

Page Composer: Creator of document files (also called a Desktop Publisher and can be the same person as the Designer). Handles typography, formats, assembles all the elements into page form, works with the Designer to carry out design, and works with the Print Group on execution. Traditionally, this person would have been several people, such as Typesetter, Keyliner, and Production Artist. These functions are combined in the electronic process.

Communication between professionals cannot be stressed enough. As software and hardware evolve, their capabilities become more and more powerful. With each new release, there is more for the professional to learn. Success today will reward those who can learn the fastest, not necessarily those who learn with the most depth. And because it is impossible for any one person to know everything in the process, learning from one another will enhance both individual potentials and enhance the process itself.

PRINT GROUP

The Print Group comprises the high-resolution service and provides finished publications. Publishing is a custom-manufacturing process. Technological advances will continue to streamline this process, but these professionals will still be needed.

Buyer: Whomever purchases reproduction capabilities. This can be any of several professionals (so this one term covers them all). For example, the Client Project Manager, the Designer, or the Page Composer can be the customer of the Imaging Center. The Client, a Print Buyer, the Designer, the Page Composer, or the Service Bureau can be the customer of the Printer.

Most Service Bureaus grew out of the typesetting industry and have a strong typographical and black and white imaging background. Most Trade Shops evolved from the color separation industry and have a strong background in color. Today they are blending together as each takes on the other's capabilities; the composite term is Imaging Centers.

Imaging Center: A merging of prepress operations. This term encompasses the functions of both the Service Bureau and the Trade Shop. The Service Bureau has imagesetting equipment for outputting camera-ready paper or plate-ready film—the Page Composer dictates any color separation instructions. The Trade Shop (or Color Separator), handles all color separation and film preparation for plate-ready output—the color separation instructions are handled within the Trade Shop.

Production Coordinator is the contact person within the Imaging Center who interacts with the Buyer. This person receives the files and makes sure they are properly prepared for imaging.

Prepress Specialist receives the document on a computer disk and readies the file for output. The Prepress Specialist handles the output through an appropriate technological device to match the specifications of the Buyer for the printer. Traditionally, these professionals are scanner operators, camera operators, and strippers. These functions are now coming together in the form of prepress operations on the computer, controlled by sophisticated software.

Printer: Advises on output, receives film and proofs, handles and supervises all printing through delivery.

Plate Maker receives film and exposes plates according to the specifications of the printing press.

Press Operator runs the presses, balances the color to match the proof, and monitors quality control.

Binder completes the manufacturing process by taking paper sheets or rolls from the press and converting (through cutting and folding) into the final form, such as books, magazines, or stationery.

Shipper takes skids or boxes of the printed publication from the dock of the printer and delivers them to destination(s) selected by the Client.

Within this chain of professionals (each of whom may subcontract portions of the work), responsibility shifts through the process, but always remains under the Art Director's purview, and subject to the Originating Client's final approval.

VIEWPOINTS:

ROLES IN TRANSITION

“ Good work has rarely come out of an adversarial relationship: it's important to abandon relationships that are a no-win. A designer needs to be generous with clients, to accept responsibility for a good relationship, but also to be demanding that the client come forward and meet them half-way.

The majority of the work we do is difficult, not because our clients are incapable of being good clients. They may be educated or uneducated and that's not what really matters as much as whether they enjoy the process and trust us. Those with whom we have a cooperative working relationship can help get through any problems. All of our clients are trouble because they're all parts of organizations with time schedules and budgets—design is a really difficult business. Clients never have enough time, never have enough money, their products are often being introduced before they're ready, they're rescheduling, they often can't get products to you for the scheduled photo shoots or they change the name of the product after the photos are shot but the direct mail is already committed so the deadline doesn't change. All clients have these problems. The ones we choose to stick with are great. They're really nice people, they care about their products and their businesses. We love their work because it's creative and interesting, so everyone stays up late, works weekends, and tries to do it.

If the client cares, there is always a way to resolve problems. Even our best clients don't always love us and they *always* think they're paying too much. For what they're paying, they think you should be able to answer all their problems. I understand this because I feel this way about my own staff! I'm very sympathetic to the client. ”

Nancye Green, designer
Donovan and Green
New York City

“The challenge for the designer in the '90s is to create the opportunity for good design. Before you can begin to put pencil to paper or mouse to pad, you have to evaluate, educate, and take control of the circumstances in order to create an opportunity that is conducive to good design! This means that designers, now more than ever, have to approach every project from a number of angles. You have to be the businessman that puts the project into financial perspective for the client. You have to be the research department that evaluates new processes and techniques. You have to be the educator who explains the new processes and mediums available. You have to be a CFO to figure out the client's budget and proportion of moneys to allow for good design. And if you do all of these things well, then you've earned the right to be the designer that knows how to deliver a great design.”

John Brady, designer
John Brady Design Consultants
Pittsburgh, Pennsylvania

“The design field that I entered forty-odd years ago has gone through a series of profound changes in the past five years. Without being a social historian, one assumes that these changes have been generated, in no particular order, by the recession, the dramatic overpopulation of the profession (with no respite in sight), the extraordinary influence of the computer, and the more cosmic issue of living in a time where the erosion of historical beliefs and values produces confusion and alienation. It is hard to describe the overall spirit of this moment within the field, but confused and ungenerous come close.”

Milton Glaser, designer
Milton Glaser Inc.
New York City

SHIFTING OF ROLES 2

New tools inspire people to do what they do more effectively. Deeper change, such as new ways of thinking and new working habits—in other words, cultural changes—come more slowly. Changes that inspire new habits are conscious, such as transforming into a paperless office or using new media to communicate in new ways. These changes of habits evolve slowly with the human capacity to absorb. The unconscious changes, such as the transformation of an industry, can happen with surprising speed. These changes sneak up on people who may find themselves suddenly out-of-date. More than affecting individuals, electronic publishing is causing all communication professionals to redefine both how they work and who will do the work. Major changes transforming publishing include:

- production becomes more automated and old skills become obsolete.
- management may perceive that fees for design and production should be less due to technology.
- students spend significant portions of their credit hours learning software
- the lure of technology has let to a greater influx of graduating design majors per capita.
- career-shifters find the ease of use and portability of electronic publishing attractive.
- technology is becoming easier to learn, placing capabilities in the hands of more workers and enabling more to publish than before.
- roles between professionals are blending and growing less definable.

Some publishing experts see electronic publishing as an industry in crisis: fees are dropping and competition is increasing. Supply is greater than demand. Although the demand keeps growing for publishing, there are more ways to achieve it. In many cases, electronic publishing actually is not cheaper or faster than traditional methods, but the industry mythology perpetuates such expectations. What doesn't change are the reasons for communication: the marketing goals of the Client Group. These goals can be achieved in more ways today than in the past. To reach these goals requires clear vision and the appropriate use of talents, skills, and resources. With a variety of meals on a menu, choosing the most nutritious meal makes health sense. With a variety of media and methods, choosing the most appropriate and targeted medium makes business sense.

MAJOR DECISIONS
plan project

ORIGINATING CLIENT GROUP

In pre-electronic publishing times, the publishing process used to hum along like a well-oiled machine. The writers wrote, the designers designed, the typesetters keystroked, the keyliners pasted, and the printers made film and plates and then printed. The Client was in a better position to remain detached. As technology has made the roles more ambiguous, the Client needs to show more leadership and knowledge to best manage the process. Decisions that the Client needs to control include:

- defining who is to do the work and what their roles are.
- articulating the expectations of the process such as speed and cost goals.
- setting up the process for ongoing work or for series that build.
- making sure that staff or vendors are knowledgeable and have the resources to best get the job done.
- balancing technological considerations with what is needed to reach goals—the ability to make decisions without being hands-on but understanding the issues and capabilities.
- utilizing resources to keep current and evolve plans.
- being willing to listen to professional sources and learn new ways of solving challenges.

CREATIVE GROUP

Conceptual and artistic skills must balance with technological skills in the creative professional. The further away from idea generation the work grows, the more technical expertise is required. However, to fully take advantage of new media such as interactive presentations or multimedia publications, the creative professional must understand the new visual language opportunities afforded by technology. The publishing artist must:

- grasp the differences between media and know what skills they need to expand.
- understand what business they are in and try not to be everything to everybody.
- understand how to use the tools for goal-purposes versus for mere technique purposes.
- have the ability to choose appropriate tools.
- be able to explain the process to decision-makers to build understanding and collaboration.
- identify the level of design that is appropriate to the assignment (see sidebar).

PRINT GROUP

Technical expertise comes in definable levels. But technical expertise alone turns the products of skills into commodities. Those professionals in the production side of publishing know that service skills, business connections, and business mastery are needed to remain competitive. Hardware and software are easy to buy—especially as cheaper desktop systems replace the high-end digital imaging devices of the past. To manage in the spectrum of technological possibilities requires that the print professional:

- have clear business definitions and markets.
- emphasize customer service and responsiveness in addition to technical expertise.
- retain a visionary technician who can keep capabilities on the forefront.
- have management vision that can choose appropriate technology for business goals.
- maintain the ability to change quickly in response to customer needs.
- possess the ability to compress the process and turn work around faster than in the past.

Levels of design

Designers accomplish marketing goals using their talent, training, evolved skills, and depth of experience. Any publishing project requires certain levels of creativity, planning, and execution. From the simplest to the most complex, the different levels of design services are:

- *knowledge and use of typography*—organizing information, formatting text, utilizing templates, and producing documents. Requires training not only in software but in the rules of typography.
- *format, composition, and readability*—arranging and composing the words rather than simply placing them on the page—leading the reader into the text through effective use of type faces and selections.
- *pagination and sequencing information*—planning the text through multiple pages and determining what should be on each page; sensitivity to making information easy to read through use of composition and sensitivity to typographic crafting.
- *overall visual scheme*—giving pages more distinction and impact by adding visual devices such as illustrations, borders, photographs, or other graphic elements.
- *graphic theme*—finding visual ways to express the content, rather than having the reader rely on words alone—in other words, a visual language.
- *expansive programs or campaigns*—developing flexible themes that give visual cohesion, recognizability, and consistency—in other words, an overall image that can be applied to a marketing vision as expressed through many pieces.

The first three categories involve layout, the arrangement of elements. The last three involve the conceptual underpinning which conveys information through visual language. Marketing purpose and client goals dictate the appropriate level of design.

It is important to recognize the distinction between the computer as a design tool and the computer as a production tool. The computer makes most production tasks faster than traditional methods. But as a design tool, the computer does not compress design time; the tools might enhance the execution of ideas, but generally (except for comping type and patterns), it takes the same or more time than traditional methods. Planning helps ensure the best use of computers as design and presentation tools.

VIEWPOINTS:

TECHNOLOGY AND PURPOSE

“Basically we have two kinds of projects. We need designers to help us sell generic services and to help us sell specific services. Generic services include things like our accounting firm's financial capability in tax practices or in telecommunications consulting. Specific services include things like providing an audit for a particular company, which we call an 'engagement.' The design projects that sell generic services are likely to be brochures, newsletters, or an identity. When we are going after a specific project, designers are more likely to be called in to design a presentation format, to help design charts, or to do a lot of nitty-gritty stuff that's going to clarify our proposal to get that engagement. In that way, we'll use design to really help with our communication process, with new prospects. As far as I'm concerned, that's where the rubber hits the road. For instance, when you have a new business presentation, and it looks like hell, you give it to a designer, who looks at it and says, 'Okay, we need sections and organization, and let's do this and that.' He sets it up on his desktop publishing system and all of a sudden it becomes something very unique and appealing. When we are competing for a significant piece of business, we will do whatever it takes to position our firm successfully against the competition.”

Robert Moulthrop
Vice President
Scudder, Stevens, & Clark
New York formerly
marketing director
KPGM Peat Marwick

“Almost everyone, up through the highest ranks of professionals, will feel increased pressure to specialize, or at least to package himself or herself as a marketable portfolio skills. Executives and what used to be called managers, will have undergone probably the most radical rethinking of their roles.

More and more of the population will be caught up in the defining activity of the age: scrambling. Scrambling for footing on a shifting corporate landscape—cynics will call it a free-lance economy—where market forces have supplanted older, more comfortable employment arrangements. Scrambling to upgrade their software, their learning, their financial reserves. Scrambling even to carve out moments of tranquillity under a banner blazoned FIGHT STRESS, a banner flapping like a Tibetan prayer flag in the gales of change.”

Walter Kiechel III
Managing Editor
Fortune magazine
New York City

“ Desktop publishing has clearly facilitated our editorial and business operations. The flip side is that it has reduced interpersonal interaction. It still takes people to do layout, design, and write text. What the PC has done is facilitate getting the message from the brain to the page. Has it improved the editorial product? No. The quality of our various magazines is still in the minds of the creators.

Gilbert Grosvenor
President and Chairman
National Geographic Society
Washington, DC

“ Someone with no background in journalism, design, or production should be put in a publications position. That's lousy management. Why does our culture entrust communicating to untrained employees? Would you have your teeth drilled by an untrained dentist? Very few even risk a student barber, whose mistakes grow out in a couple of weeks. Yet we assume that organizing thoughts and transmitting information vital to the company's—and the recipient's—interests can be done by a beginner who admits to being a know-nothing. We deserve what we get.

Bosses need to understand that no machine can substitute for professional judgment and experience. Communication is a highly sophisticated and specialized activity, demanding skills that take time to learn and hone. Just because someone can write a letter does not mean they can become a skilled newsletter editor overnight, no matter what the software makers claimed. ”

Jan White
designer, author, consultant
Westport, Connecticut

“ Graphic design is a silly title that has little or no meaning these days. Just about anyone can design a newsletter. Everyone can get fonts, everyone can get charcoal drawings with a software plug-in filter applied onto a photograph. The things that used to differentiate us from other professions and businesses have turned into everyday commodity products. In a sense the computer is a great equalizer. Why should someone hire us to design a newsletter when they can buy this off the shelf? This is a difficult question; the answer, I believe is that we need to provide added value and experience that cannot be bought 'off the shelf.'

Clement Mok
Clement Mok Designs
San Francisco CA

3 SECRET OF SUCCESS: ALWAYS PLAN AHEAD

ADEPT (Association for the Development of Electronic Publishing Technique) represents all professionals who work with desktop publishing to create communications. Members include those who utilize computers as management and planning tools, and those who create and produce documents on computers. It is an educational forum where users can go for exchanging information. For information: ADEPT, 2722 Merrilee Drive, Fairfax, Virginia, 22031, 1-800/ADEPT11.

Barbara Golden of Computing Solutions in Chicago, an early adopter of electronic publishing, finds that all disputes within the graphic arts industry seem to fall within five areas of potential conflict. She recommends that you always ask yourself these questions as you proceed from one phase to the next:

1. Do we have the **training** and skills for this project, or are we venturing into the unknown and need to factor in experimental time?
2. Does our **planning** provide for an organized approval process, define decision-making set-up, project management, and assign appropriate responsibilities?
3. Are we **communicating** our expectations and budgetary considerations to everyone involved in the process?
4. Are all the right professionals **participating** together early enough in the process to receive input for planning, both inside and outside the organization?
5. Are the groups **cooperating** with one another? Are the participants open to suggestions and learning? Do they know what questions to ask?

These are ingredients for a successful project. Use this list both at the beginning to foresee what might derail plans, and, later, to discover what went wrong if something does.

Planning also involves understanding responsibilities and accountability. As the process changes and roles evolve, questions of who is responsible for what may continue to transform. However, each person involved in the publishing process should always be prepared to accept responsibility in three major ways, no matter what their role:

- Prevent potential difficulties:
  - plan—set up an efficient procedure.
  - communicate—to all those in the process.
  - define expectations—to each participant.
- Control quality:
  - explain—instructions to producers.
  - monitor—progress, expectations, and excellence.
  - report—to decision-makers.
- Facilitate the work flow:
  - handle the approval process—make sure progress is complete and occurs before the next step begins.

▻ deal with unexpected results—allow time for experimenting.

▻ control the budget—throughout the project to ensure that there are no surprises at the end.

ORIGINATING CLIENT GROUP

MAJOR DECISIONS
plan project

It is worthwhile at the beginning to take the time to systematically plan how the project will be administered. Careful planning can save considerable time later.

CREATIVE GROUP

Clear communication throughout the process is the key to success, both with the Client Group and the Print Group.

- Expectations need to be spelled out regarding quality, timing, and what happens at each phase.
- The Art Director, as the manager of the Creative Group, has responsibility for the project process, and
  - must understand the requirements of electronic procedures—in addition to traditional processes—to know what procedures are most appropriate.
  - monitors decision points as the work progresses.
  - makes sure the Client approves key phases and thus is ultimately responsible for content and accuracy.
  - is responsible for advancing to each next step depending on "who has the ball" at each point in the process.

PRINT GROUP

Having a clear order form and procedural information available to customers can help facilitate an efficient process. Bear in mind that these instructions need to be simple and concise.

PROJECT MANAGERS CHECKLIST

- Meet with decision-makers and check completeness of in-coming element
- Quote project elements
  - > get printing bids
  - > estimate design and visual elements
  - > estimate production fees
- Receive project
  - > schedule work flow
  - > meet with Creative Group and give assignment
- Concept development
  - > monitor project progress
  - > review designs
  - > check budget against design
- Presentation
  - > present to all decision-makers with designer
  - > mention any appropriate budget considerations
- Finished art development
  - > monitor project progress
  - > check before final lasers
  - > supervise approval process and receive alterations
  - > receive Client approval
  - > review budget
- Final output
  - > check to make sure complete
  - > review color specifications
  - > check proofs then negatives
- Printing supervision
  - > receive and review printing proofs
  - > route proofs to Client and obtain signature
  - > communicate costs on alterations
  - > press-check project
  - > receive samples
- Analysis and billing
  - > review actual budget
  - > check on delivery
  - > check with Client to see if happy
  - > review billing information

VIEWPOINTS:

APPROPRIATENESS

" There is a tendency today for companies to go directly to a design firm for graphics and for design firms to go directly to companies before an overall marketing communications plan has been developed. This results in wasted time and money for both the design firms and the companies. Design flows from corporate objectives, and until the objectives are established, a design cannot be effectively developed. Design should be a part of a company's overall communications plan. Thus, design firms need to strategically align themselves with those who develop the communications plan to ensure they will be considered for the design work. On the other side of the coin, companies, should be sure to develop a corporate communications program first, before requesting design services. "

George Sepetys
Comark Group
Detroit, Michigan

" The importance of communication in business is by now beyond dispute. Perhaps it always has been; but never before has communication been so widely acknowledged as one of the most significant functions of corporate life. There is almost universal agreement that the ability to communicate is essential both to the success of careers within a corporation and to the success of the corporation itself. "

Ralph Caplan, communications consultant and writer
New York City

" If clients don't plan properly, they may end up with a beautiful piece, a style they like, but without the appropriate content. They may find that they are riding a beautiful camel when what they really need is a horse. "

Jim Madden, president
Rider Dickerson, Inc.
Chicago, Illinois

“ Each successful company begins by asking the right questions. The right questions are not: *What kind of image do we want to project?* and *What look would we like to have?* The right questions are: *Who are we? What are we like? What have we got to sell? What have we got to say?* The answers to such questions become the bases for communication programs that are entirely appropriate to the companies they represent. ”

Ralph Caplan, communications consultant and writer
New York City

“ As a professional consultant, the graphic designer can determine the feasibility of a project by incorporating his or her knowledge of the technical resources available. Often clients choose to develop projects and *then* bring in the designer. This can be an inefficient use of the designer's capabilities, since many decisions may already have been made that the designer should have been consulted on. The result can be unnecessary delays, additional costs, and inadequate design conclusions. The sooner the designer is called in to consult on a project, the easier it is for the designer to help steer the project to the best graphic solution. ”

The Graphic Artists Guild
Pricing and Ethical Guidelines

“ Good ideas often materialize as the client and designer work together in the planning phases. Everyone involved with the project should freely express their needs and goals. The more a designer understands the intent of the project the better the end result. If the designer is brought in too late, there is not enough integration between content and visual expression. He is merely spreading design frosting on a cake of information, which is an unrewarding way to spend promotional dollars. ”

Michael Waitsman, designer
Synthesis Concepts, Inc.
Chicago Illinois

4 ESTIMATES, QUOTATIONS, AND PROPOSALS

For further reference, many associations concerned with electronic publishing are included in the Appendix.

International Association of Business Communicators (IABC), One Hallidie Plaza, Suite 600, San Francisco, California, 94102, 415/433-3400.

Before a project begins, a financial arrangement should be put in writing. It should specify the scope of work, parameters, fees, and terms. This is usually based on an estimate, a quote, or a proposal. The purpose of the financial arrangement is to establish price and to predict cost. Although these three words are often used interchangeably, each has its own meaning.

The non-binding, though, can later be hard to enforce. There is no such thing as a "ball-park." Once a figure is named, it sets the tone and is remembered. Also, it is very easy to underbid when an estimate is given spontaneously such as in an initial discussion.

- An **estimate** is a non-binding preliminary projection of project costs.
- A **quotation** is a firm statement of price for specified work to be performed. It is often subject to credit approval.
- A **proposal**
  - outlines the intent of the project (such as market and purpose).
  - specifies project parameters (such as direction, elements, and timing).
  - includes estimates for design and production fees.

Estimates, quotations, and proposals are usually prepared without charge. However, fees may be appropriate for extensive proposals. If so, this should be agreed upon in advance.

In special cases—for small projects or for long-established business relationships—less formal means, such as verbal quotations or simple hourly billing, are preferable. At the other end of the spectrum are business arrangements based on contracts. In general, it's easiest to resolve any problems when the agreement is in some form of writing, but appropriateness should be the main guide.

MAJOR DECISIONS

plan project

solicit proposals

choose Creative Group

ORIGINATING CLIENT ROLE

For budgeting and scheduling, the Client needs an estimate, quotation, or proposal. The Client then:

- receives all necessary approvals within its organization to begin the project.
- defines the internal decision-makers and approval process.
- identifies any trade secrets or proprietary material they make available to the Creative Group. This may need to be covered by a nondisclosure agreement with the Creative Group in order to be considered confidential.

- gives the Creative Group the go-ahead to proceed with the project, often secured with advance payment, purchase order, contract, or other appropriate means.
- negotiates ownership of project materials.

CREATIVE ROLE

The Creative Group may bid on a project with an estimate, quotation, or proposal, depending on the preferences of the Client and the Creative Group. After they have developed and presented the bid to the Client, estimates, quotations, and proposals not accepted by the Client within thirty days may need to be updated with revised procedures or fees. The Creative Group:

- solidifies the fees, the schedule, and the financial agreement before beginning the project.
- often provides an outline of the project process and schedule.
- explains the process and discusses expectations with the Client.
- where appropriate, the Art Director signs a nondisclosure agreement, which
  - binds the entire Creative and Print Groups to confidentiality concerning the project.
  - can be used to communicate these needs to the entire Creative and Print Groups.

Example of a commonly overlooked expectation: Many Designers charge extra for editorial changes, and the Client is surprised at the fees after the work is done. To avoid misunderstandings, the Creative Group should clarify extra fees in advance and perhaps recommend a contingency for alterations that may or may not be needed.

PRINT ROLE

Print services rarely provide proposals. Rather, they bid on work with estimates or quotations. Estimates or quotations not accepted by the Buyer within thirty days may be subject to revision.

- An estimate needs to be revised if the information or specifications change.
- A quotation
  - is based on the exact materials, costs, and job requirements specified.
  - should indicate appropriate taxes (such as on film and materials), deliveries, and any other extras that will be charged. In other words, a quotation should give the Buyer a clear idea of the future final invoice.
  - must be in writing.

VIEWPOINTS:

RESTRICTIONS

"Designers and clients must be responsible for communication and understanding each other. The smallest and most insignificant detail can often be the one which causes the machine to break down. When it comes to working agreements, you need to work to agree. Be clear and complete."

Richardson or Richardson
Hopper Paper Company

"Budgetary constraints should be viewed as basic problematic parameters and not loathed as obstacles to a project's full potential. This is difficult, but circumstances, such as these budget-conscious days, force a designer to be more resourceful and to make each specification work more, and hence worth more. With each aspect of a design scrutinized to be most effective, less expensive, materials and processes should look better. In these times, designers must continue to innovate. The default option to use what has worked before or to arbitrarily revive precedent should be avoided. Clients and consumers will be using new gauges in criticizing items as appropriate or inappropriate. They will be inclined to call their new perceptions 'practicality' or 'pragmatism' when it is simply the next shift in aesthetic taste. We should remember not to cater to these changing perceptions, but as designers, change the perceptions themselves."

Terence Leong
Context International
New York City

"Neither estimates or budgets are predictive. They will not tell you what is going to happen. Their principal value to the client is to set a limit on costs. The principal value to the designer is to provide a guide to help them keep costs under control. Even the most carefully constructed estimate will not help you if you allow a project to get out of control."

W. Daniel Wefler, publisher
Wefler & Associates, Inc.
Evanston, Illinois

“The single most important ingredient in estimating is prior experience. Design estimating is more difficult than manufacturing estimates because there are more undefined elements. Defining them is part of the design process. Design firm estimating will be dependent upon previous experience with similar work. Experience helps you understand the time needed, the problem potential, the quality factor, and the price history of what clients have paid.”

W. Daniel Wefler, publisher
Wefler & Associates, Inc.
Evanston, Illinois

“We saw five designers' presentations and asked for three proposals. We were looking for a designer who projected excitement, visually and verbally. We get excited about what we do, and we like to see that kind of excitement from a designer. Then, we asked to see some numbers. As a matter of fact, the designer we chose was a little bit higher than the others. So it wasn't straight dollars; it was the rapport established in the initial meeting that made us comfortable.”

John and Bill Schwartz
owners of Schwartz Brothers
Restaurants, Washington state

“I want a proposal to show me that a designer understands the project objectives; what the problem is that this particular project is going to solve. It can be structured in any way that truly accomplishes this.

The purpose of a proposal is to make sure there are no surprises! You may want to consider it as a kind of discussion draft. If it weren't that, it would be a contract. A designer must reassure me that they understand what I said in our initial meeting. They must also make sure that I am clear about the specific services they are going to provide.”

Robert Moulthrop
Vice President
Scudder, Stevens, & Clark
New York, formerly
marketing director
KPGM Peat Marwick

5 FEES FOR PUBLISHING PROJECTS

Fees anticipate the costs for which all work will be performed and billed. This includes hourly rates, per-project rates, and monthly retainers, etc. Clearly communicating the financial aspects of a project early helps to determine the scope of the work and can avoid potential misunderstanding. It leaves the publishing team free to concentrate on the work itself during the process and sets up a good collaboration.

MAJOR DECISIONS

plan project
solicit proposals
choose Creative Group
finalize the proposal

There are no standard fees in the graphic arts because no two projects and no two suppliers are ever the same. This increases the importance of clear agreements.

ORIGINATING CLIENT ROLE

In planning and preparing for a publishing project, the Client defines parameters that include the following:

- goals—what the project needs to achieve.
- scope—size, range of market, and positioning.
- intent—the audience the publication should reach.
- direction—guidance regarding appropriate content and style.
- background—previous information, history, industry information, etc.
- budget—resources available.
- time-frame—projected completion date.

The Client also verifies fees with the Creative Group and approves all costs before work begins.

The *Graphic Artists Guild Pricing and Ethical Guidelines* is a good source for finding out what suppliers charge. For information: Graphic Artists Guild, National Office, 11 W. 20th Street, 8th Floor, New York, New York, 10011, 212/463-7730.

CREATIVE ROLE

Fees have many components and each organization has their own method for arranging them. Generally, in-house departments conform to company policy when charging other departments for services.

Outside vendors, such as design firms and freelancers, need to be compensated fairly, depending on their circumstances. Possibilities for fee arrangements may be the following:

- On a per-project basis, including
  - time
  - materials.

- On a payment plan, such as
  - 1/3 in advance of the project
  - 1/3 upon approval of design
  - 1/3 upon completion of art work.
- On retainer, which is a monthly billing arrangement guaranteeing a consistent payment per month.

A Client may prefer a retainer method of billing with design firms they use regularly to plan their costs over a long period of time. This can enable smooth budgeting within a defined time-frame.

The Creative Group bases fees on the following considerations:

- Size and scope of project
- Uses to which the materials will be put (the more extensive the use, the greater the compensation—(see Chapter 28, page 159)
- Complexity of parameters and difficulty of assignment
- Nature of Client Organization
- Project and account management:
  - decision-making processes
  - presentation processes
  - necessary documentation
- Technological capabilities.

There are several cases where design firms may provide lower fees than they normally charge: for start-up organizations, nonprofit causes, markets they have not worked in before, or personal connections.

The Art Director provides a project description (usually in the form of a proposal), which may include the following:

- Time-frame
- Estimate of outside costs
- Itemization of fees
- Payment arrangement that fits the business practices of both the Client and the Creative Group.

PRINT ROLE

Charges for work may be based on the following:

- Job quotation.
- Hourly rate.
- Unit price.

The project is billed as work is completed:

- Estimates should be provided in writing to the Buyer before work is performed.
- Charges are considered accepted and due unless the Buyer contests in writing within 15 days of receipt of invoice.

VIEWPOINTS:

COMPENSATION

" The most general principle for determining the price of artwork is that the price should be in relationship to the value of the intended *use* the buyer will make of the art. This means the more extensive the use, the greater the compensation to the artist. Some inexperienced art buyers are shocked by such a concept. They assume that they are buying a *product* at one flat price, with which they can do whatever they wish upon payment. But artists normally sell only certain *rights* to the use of their creative work. The greater such rights, the greater the compensation required. "

Pricing and Ethical Guidelines, Graphic Artists Guild Handbook

" There is no orderly market for design services. There is no formal mechanism to collect and transmit information about design prices. No stock market quotations or price indexes. Design is bought and sold in thousands of negotiated transactions that are only loosely related. Buyers can seek competitive proposals and get a feel for pricing. Designers must work a little harder seeking to find out what their competitors charge for the same work "

W. Daniel Wefler, publisher
Wefler & Associates, Inc.
Evanston, Illinois

" Anyone researching design fees will quickly learn that although there are several published guides for calculating graphic design or desktop publishing hourly rates, very few sources for hard data on pricing exist. Since there can never be an absolute set of national fee standards due to federal trade restrictions, the best sources of information continue to be informal AIGA peer group networks and the 'Graphic Artists Guild Handbook of Pricing & Ethical Guidelines'. "

Juanita Dugdale
design correspondent and contributor to *HOW* magazine
Hastings-on-Hudson, NY

"Excellence is not necessarily a function of budget. It's a function of attitude. When budgets are low, that's the perfect time to propose something a little bit different. And I think it's *always* important for designers to look for the opportunity to suggest an approach that goes beyond the usual. . . . If you don't have a huge budget, you have to leverage, and leverage comes from creativity on the part of the designer."

Edwin Simon, president
The Pelican Group, Inc.
formerly vice president of
Sikorsky Aircraft, division of
United Technologies
Hartford, Connecticut

"Designers who now work in the challenging new fields of interface product development and online service design are pioneers in the unknown territory of electronic pricing. They must rely on judgment alone since published dialogues about pricing have yet to appear and intense competition among a few select consultants keeps rates a closely guarded secret. A further complication: choosing fees or royalties as the appropriate form of compensation."

Juanita Dugdale
design correspondent and
contributor to *HOW* magazine
Hastings-on-Hudson, NY

"The only way to sell quality in a budget-conscious world is to show the power of an idea. It stands on its own without help from trendiness. Creating an image with impact means having to examine the roots of our needs and perceptions. Powerful design performs well, but also hits an emotional chord. Ultimately, quality design will make us all successful. Satisfying business needs is why we're here—giving them the best is what makes us valuable even when budgets are tight."

Mike Scricco
Keiler & Company
Farmington, Connecticut

"I believe the biggest challenges facing the design industry involve working with greatly reduced client budgets, maintaining your own profitability as a design firm, and still delivering high quality work. The industry has come a long way in the past eight to ten years. Fortunately, it is finally being recognized for its tremendous worth to business. Although business in general has become increasingly more competitive in the 90s and clients are pressuring us even more in terms of price, we can only be a detriment to ourselves by making price, as opposed to quality and value, the primary issue as the basis for being awarded projects."

Mary F. Pisarkiewicz, designer,
Pisarkiewicz & Company
New York City

6 SPECULATIVE WORK AND FORMATS

Several materials may be requested by the Client to support the proposal without being charged a fee:

- Blank formats (or "dummies") showing paper stock and weight, size, and bindery, are created to provide the Client with a "feel" of the project. These function as production guides. This work, when provided, is generally accepted as a fair business practice.
- Speculative (or "spec") design work, in the form of laid-out or designed formats (or "comps"), is produced by the Creative Group for a Client's review—in expectation of being awarded the project. This is not recognized as a fair business practice.

ORIGINATING CLIENT ROLE

MAJOR DECISIONS

plan project
solicit proposals
choose Creative Group
finalize the proposal

The Client may request blank formats to help illuminate the proposal. Whether the Creative Group charges the Client for these depends on the extent of the creative work involved. The Client may legitimately ask the Creative Group to include:

- preliminary items needed to form a printing budget, such as paper formats and blank samples demonstrating printing specifications.
- information that is particularly important if new processes (new to either the Client or to the Creative Group) are to be incorporated as part of the project, such as examples of new typefaces, new software application combinations, unusual illustrative or photographic treatments, new printing techniques, etc.

ADEPT joins all other professional publishing organizations in supporting the position that creative work should not be done on speculation. Because this is the least tangible of the publishing phases, because it requires the greatest amount of market understanding, and because it is the basis for all other fees, it is the spark that sets the other processes in motion. This service should not be under-valued or given away.

Some Clients may need an explanation of why speculative creative work should not be performed.

- The project parameters may not be completely formulated.
  - There are very few documented cases where spec design works out favorably without the in-depth commitment of a Client/Designer collaboration, therefore the Client is not comparing concepts developed from equal information.
  - Design firms creating spec work may base concepts on different approaches and knowledge than their competitors.

- Spec work will not receive a Design Group's best efforts.
  - The Creative Group can't afford to spend as much time as they would if they were getting paid.
  - The Creative Group will often not have an in-depth understanding of the assignment.
  - One Creative Group may be competing against another, which means that one or more Groups will not get paid for their work. This is the purpose of portfolio reviews, proposals, and interviews as a way of differentiating one firm from another.
  - This is the most in-depth creative thinking of the Creative Group, and therefore, worth the most financially.

CREATIVE ROLE

Supplying blank formats can focus a proposal and help a Client to visualize the size and scope of the project. Blank formats:

- demonstrate paper samples
- show arrangement of components
- remain the property of the Creative Group.

Speculative designs and concepts are not to be provided without a fee payment.

- Speculative design is considered to be an unfair business practice. It is providing a valuable service without compensation.
- The only Creative Groups that are willing to do this:
  - wish to get a foot in the door with a potential Client
  - want to build their portfolios
  - desperately need business
  - are willing to donate their services.

PRINT ROLE

Blank formats for paper, size, or printing technique may be provided, usually by the Printer, at the request of a Buyer. However, with new electronic software that configures projects for printing, Imaging Centers will also be providing this service.

- The Print Group rarely charges for these, unless extensive time and costs are involved.
- Blank formats are not to be used by the Buyer without permission from the Print Group and without mutually agreed payment.

The Director of Publications of a famous museum asked three design firms to submit a speculative brochure cover with their proposal for a new promotional brochure.

Firm One had great experience and ability. Although their concept was excellent, they didn't have enough time to execute the presentation materials well.

Firm Two had more time to devote to the project, and their presentation was very polished. Yet their concept was not quite right. They did not realize they needed to do more research.

Firm Three was the least experienced. They attacked the assignment with great enthusiasm and spent a great deal of time on it. But their ideas lacked conceptual depth and were merely decorative.

The presentations that the Director of Publications received were all in different forms and styles. The committee had to choose between apples, oranges, and pears. It is very difficult to look beyond the presentation styles to see the best concept.

The committee ended up choosing Firm Two's work. But because their concept was not right, they had to start all over again. All the design and presentation work they did on speculation was not used or compensated for. The other two firms were not compensated at all.

Buying design this way is much like paving a road with no foundation underneath. When a Client buys creative services speculatively, they do not allow the Creative Group to do its best work.

The correct way to hire a Designer or firm is to objectively evaluate their finished work in its approach, diversity, sophistication, and completeness. The design firm's proposal shows their understanding of the project. And the final important factor—rapport—is clear from the personal interactions.

VIEWPOINTS:

INTEGRITY

"In large-scale advertising, providing "free" creative solutions in a competitive format is commonplace. In brain surgery, it isn't. Design must be somewhere in between. Clients should determine whether they're purchasing a design, or hiring a designer.

Designers need to remain open-minded. That's what makes them valuable in the first place. Are there innovative routes which can meet everyone's needs? Perhaps.

At all costs, the design process has to be valued. Without value, it doesn't matter a hill of beans whether it's done on speculation or paid for."

Richardson or Richardson
Hopper Paper Company

"Several major studios are now quietly developing new business by working on spec. The end result is a client who expects more for less, or for nothing at all. But more importantly, the small design 'boutiques' (which have always been the heart of industry creativity) usually count on one or two major projects as their life blood and cannot afford to work for free."

Regina Rubino, designer
Louey/Rubino Design Group
Santa Monica, California

"Good advertising is conceived in frustration and delivered in agony. It is precious stuff. It shouldn't be wasted and it shouldn't be given away free, no matter how eager an account executive or agency president is to please a client."

John E. O'Toole
President
Foote, Cone and Belding

“ In our society, we express our respect for work by paying for it. When we refuse to do so, we are expressing contempt for the work and the worker. In these difficult times, the fundamental rules of human conduct are under attack in and out of business. The only appropriate response is not to allow our own sense of values and self-respect to erode in the face of it. ”

Milton Glaser, designer
Milton Glaser Inc.
New York City

“ Anyone doing work on speculation is likely to go unpaid. But the fact that one person or firm works on speculation makes it likely that others will be asked to and feel a greater compulsion to answer such requests affirmatively. So the issue of right and wrong cannot be decided on the basis of the individual alone. While work on speculation is likely to damage those who do it, it certainly erodes the creative and financial health of the community. This is why the organizations representing the creative community have drawn codes to regulate fair practices, and why these codes take a strong stand against work on speculation.

”

Tad Crawford
writer and publisher
Allworth Press
New York City

7 TRAINING

Although managers understand that the technology is only as good as the person using it, many don't seem to believe it. With the promise that software gets easier and easier to learn and use, many budget-conscious executives invest less in training employees. Investing in people is investing in a resource that can quit and find another job. Even though the equipment grows obsolete, hardware and software are more tangible investments. Investing in employees is intangible, with results that may be hard to measure. However, effective training offers many benefits:

- employees will have the knowledge to choose the most cost-effective methods to handle various assignments.
- employees will be better able to help each other.
- upgrading and new technologies will be easier to absorb.
- employees will develop the habit of learning and will thrive in an environment that is conducive to expansion.
- skills learned in one software package can help workers learn other packages.
- the individual professional can stay competitive in the job market.
- workers can take responsibility for their own training as job security, increasing initiative and value to their employers.

The commitment to an electronic publishing system is not complete with the purchase of hardware and software. Adequate training is imperative, and may cost over twice as much as the initial workstation investment. Too many companies underestimate the importance and cost of training. But if staff is adequately prepared, then quality and production efficiency will be realized.

Training is an on-going commitment and inspires the ability to thrive on change. It is human nature to resist change, but the one thing in life that *is* constant is change. To stay ahead of it in an accelerating era of technological changes requires an attitude of commitment, the habit of learning, and the ability to absorb information quickly.

ORIGINATING CLIENT

The Client and the manager can create the atmosphere of learning and growing within their organizations and inspire it in the organizations that they hire as resources. A company's commitment to training reflects a commitment to excellent work. Qualified employees need technical skills. But those skills need to be continually reinforced. Many forms of training should be encouraged as part of a corporate culture. Formal training includes:

- *Classes at a training center* are an excellent way to bring employees up to speed quickly. They allow the student to ask questions and to learn along with others. But classes are only a beginning—they introduce the user to a new application. There really is no short-cut for practice, and the best way to learn is on real projects, not pretend ones.

- *An individual trainer* can be an ideal source for training. After attending a class, a user will benefit by having a consultant come in for one-half day, or for a day, to work side-by-side with the learner, working over any learning hurdles and teaching efficient techniques that apply specifically to the job.
- *Books* can augment the other training methods with supplementing more specific knowledge that applies directly to the work. Many high-quality "how-to's" are on the market.
- *Public seminars* are excellent for overviews. Their range covers business, management, particular issues, kinds of projects, specific applications, and demonstrations. No matter what other training methods are used, seminars should be encouraged because they are a good way to understand and pin-point trends.
- *Conferences* contain many seminars and provide greater in-depth study and interaction. One of the best for an overview of industry trends and changes is CONCEPPTS in Orlando, sponsored by the Graphic Arts Show Company with the collaboration of many graphic arts associations.
- *In-house* Information Center or help desk. Larger companies may have internal resources to help users learn fundamentals and maintain the system.

(For examples of informal training, see list under Creative Group below.)

Planning for effectiveness

Basic questions need to be reviewed for the Client to understand the requirements, issues, and capabilities of staff:

- Do employees know the basic packages and capabilities? (There's a difference between hands-on and overview knowledge. Some workers need both—but a manager may only need the latter.)
- Is the company willing to train—and to what extent—for specific applications?
- Is the *level* of hands-on work scaled in job descriptions? For example, newsletter desktop publishing requires different skills than four-color brochures.

Developing and implementing Page Standards within an organization that produces a quantity of work has many advantages:

- ensures that all documents are set up in the same way and that employees can cover for one another.
- accelerates the training process by showing tangible applications for projects.
- becomes a continuing reference for ongoing work.
- brings new employees up to speed quickly.
- reduces output mysteries and costs.

- Do company executives understand, and plan for, ongoing training? (The learning curve is really a learning roller-coaster.)
- Is the initiative for self-training emphasized, yet still with limits to impede wandering experimentation? (Software manuals are marvelous tools for training when workers have the time to systematically teach themselves. However, manuals generally can't take the place of some professional training and can't answer all employee questions or address job-specific challenges.)
- What specialization is expected and encouraged among employees? (Each company is a mix of expertise, for one person can't know or do everything.) How is the cumulative knowledge of staff balanced?
- Is there a plan for helping an employee adapt to the changing technology?
- Is a culture of continual evolution encouraged and are expectations, both from the technology and the employee, outlined? Are the corporate and personal benefits made known?
- Is information easy to access, such as reference materials, resources, communication with other users, as well as to the manager responsible for the system?
- Are employees encouraged to evaluate the system and give feedback to use at the next improvement stage?

CREATIVE GROUP

Few employers are willing to shoulder the expense of in-depth employee training and it often must be the employees' responsibility to invest their own time, and money, in their continuing education. Creative entrepreneurs have to factor training and learning costs into their fees and schedules, much as they need to amortize their equipment costs in their overhead. Because learning is so constant, it is hard to factor it into overhead. But it is as much a cost of doing business as paying the phone bill and the rent. There are many methods of formal training (see above list under Originating Client) as well as ways of staying up-to-speed on an informal basis:

- Learning on a need-to-know basis and having resources to meet that need (one of the best ways to learn because knowledge can immediately be applied).
- Background training (through classes, tutorials, and time spent with manuals) to learn fundamentals and capabilities.

- Intuitive experimental exploration, often achieved through making mistakes or through trying different menus and selections to see what happens.
- Asking questions of other professionals through networking, help desks, phone calls, and vendor services.
- Investigative training to increase specialization, generally through seminars, research, and trade show attendance.

The creative professional must provide an atmosphere conducive to these forms of curiosity. If the employer can provide a policy of compensation for seminar attendance and individual research, this will further motivate the employee.

PRINT GROUP

Applying the training to the production end of the publishing cycle is both easier—because its more identifiable—and more confusing—because there are so many choices. (See the above categories for training options and tips.) Possible targeted resources include:

- traveling seminars.
- conferences—local and national—that have training tracks.
- technical foundations and associations that offer courses.
- vendor-sponsored seminars that are usually free (though biased to certain equipment and software choices).
- consultants with various specialties.
- technology management firms to help with major transitioning

How to keep up and stay current:
Training is an ongoing individual commitment. Staying current helps uncover areas where more in-depth knowledge is needed. To keep up with trends and information:

- *Blend skills with other* employees or experts to complement your abilities.
- *Seek out a dedicated "guru,"* someone who is fascinated with the system and passionately expands their knowledge. They can maintain the computers, plan system expansions, keep up with upgrades and licenses, and handle daily problems.
- *Use the computer as a tool, not as a crutch.* It is important to know non-technical methods to find the best solutions.
- *Guard against over-perfection* and becoming enamored with unimportant details the final viewer will never notice (or care about even if they do).
- *Sharpen your communication skills.* How each person relates to one another, provides instructions, reports job progress, handles quality control, and works with others, becomes critical when there is less time to do things.
- *Make friends* who know things you don't for something specific.
- *Don't always do everything yourself.* For the sake of time, money, and quality, don't do what someone else can do better.
- *Identify outside resources that have the expertise you don't.* Develop relations with suppliers as a research and development source rather than shouldering the expense of being on the "bleeding edge."
- *Attend conferences and seminars* to get questions answered and interact with peers.
- *Accept that there is always more to know.*
- *Find a sense of satisfaction outside the technology*—a sense of completion from what you *do* versus *what* you have.
- *Read strategically and quickly.* Set aside specific time to skim industry information.
- *Set limits on the time you will spend on an activity.* When you start working on a project, decide when you will stop.
- *Learn to ask calculated questions* for the information you need. Success of information handling has to do with astute questions that prompt efficient sorts and overviews.
- *Determine how much detail you need* for any given project and guard against getting more information than you need.
- *Become a good editor* of the information you generate for others. Identify what reports, memos, and letters can be done away with or handled by other means.
- *Depend on electronic backup-and-retrieval* instead of hard-copy filing to save space.
- *Be careful of electronic bulletin boards.* It is easy to spend inefficient hours communicating with others (though fun to do).
- *Get away from technology* by making sure you have other activities to balance, including those that are work related.

VIEWPOINTS:

PROFESSIONAL GROWTH

“ Here are seven paradigms for integrating computers into a productive environment.

1. Proactive thinking—build a positive and empowering culture.
2. Lifelong learning—aim to learn not only the software, but also the new processes electronic publishing demands.
3. Job specific learning—implement a customized and well-planned learning track.
4. Training as an investment—partner with a reputable training and consulting firm.
5. Continuous improvement—examine workflow and continue to improve it.
6. Corporate culture of empowerment and productivity—clearly define expected results.
7. Right training for the right people at the right time—do not get confused and make training the end goal.

What can I look forward to if I incorporate these paradigms?

Full productivity will happen when the workflow is right, progress is forward, and the office is staffed with power users. When real and job-related learning occurs, so does the potential productivity result from use of the wondrous devices. Accuracy and profitability will increase, customers will be happy, morale will rise, and turnover will diminish. ”

Barbara Golden
President
Computing Solutions, Inc.
Chicago, Illinois

“ Training is *the* most cost-effective thing you can do. Don’t buy new equipment until you learn to use what you have. ”

Clint Funk
White & Associates
Northbrook, Illinois

“ The information systems professional typically views the problems of computerization in terms of the technology rather than in terms of the business. That is, when an implementation falters or fails, it is seen as a missed opportunity to use the technology.

When the technology invades every area of the business, the technician must be schooled in all areas of the business.

One of the best ways for an organization to educate its technical staff about the business is through a collaboration between technician and user during systems development. They can train each other in the business of the organization. ”

Vicki McConnell and Karl Koch
management consultants
MENTOR Group
Columbus, Ohio

“ **Great Expectations: How Users Will Behave**

Technology will be used effectively if managers know what kinds of behavior to expect from employees. Managers can use the following categories to help them set behavioral expectations:

- *Productivity*. Once started on a specific computer task, how productive is the employee in a given time period?
- *Proficiency*. How much of a specific software application has the employee mastered?
- *Efficiency*. How well can the employee use all functions mastered in a given software application, moving from one task to another with ease?
- *Effectiveness*. How well can the person choose between options in mastered software applications?
- *Attitude*. Given past experience with computers, how positive is the employee's attitude toward present and future performance with the technology?
- *Quality*. How well can the employee perform according to corporate and workgroup standards? ”

Vicki McConnell and Karl Koch
management consultants
MENTOR Group
Columbus, Ohio

“ The most expensive part of the process is fleshware—having people with knowledge and experience. ”

Chuck Weger, consultant
Alexandria, Virginia

Training Issues in the Design Studio

“The single greatest challenge for managers and designers in the age of technology is training. The only consolation is that it is a universal problem; no one is spared.

The first obstacle is getting up to speed at the same time that existing work must be completed. If the studio is already busy, with little time to spare, it is unlikely that anyone will find a *convenient* time to try something new. Many new computer setups sit idly while *real work* continues.

Each person will master the computer skills at an individual pace. Everyone should be given time to learn it outside the rush and clutter of production schedules. After an initial period of time for familiarization, real work should be assigned on the computer. This will ensure that the skills are immediately applied to the everyday work, thus integrating the computer as quickly as possible.

How one manages such training becomes the challenge. The first level is to bring everyone up to speed as quickly as possible. This should include an overview of computer concepts and applications as well as instruction on specific software. Schedule periodic follow-up sessions in the beginning so the staff is given the chance to try the tools, make mistakes, and understand how to be more efficient over a period of time. Give people time to practice and experiment. Don't assume that training will happen simply because some individuals take a course. Practice is essential; build it into the schedule and workload. Provide temporary staffing to cover for those learning the computer.

Pay for professional training, unless there is an individual who already knows the computer and is willing to take on the job of training fellow staff. Don't leave it up to each person to master the computer without the support of experienced people. The more capable and complex computer tools become, the more training we need to master them, and to understand not only how to integrate computers into the work but also associated skills, like video editing and sound.

For on-going training and to expose designers to current trends and developments in technology, make sure there is as much exposure to new ideas, products, and work as possible. Periodicals and conferences will help. Bringing in demos to the company by vendors also will open up ideas. Have the training continue after a period of time to bring to the staff industry changes and new ideas. Because you are in the business to sell creativity, new ideas will come from seeing new tools and applications.”

Alyce Kaprow
computer graphics consultant
Newton, Massachusetts

“ Responding to change doesn't just happen. We need to train ourselves to change... and train in several areas:

1. **technology training** equips one to deal with incredible constant change in how we do our work.
2. **business training** means more now than ever before. It's no longer just application (that's the easy part). We must continue to train in our changing businesses to survive.
3. **environment training** means knowing how technology and business are changing in a particular site or condition... and adapting.
4. **technique training** means studying and implementing "what works with what, when."
5. **survival training** is knowing how to deal with your own and others' change. It means asking questions about what to we hold on to, throw away, and add to cope with chaos. This means training yourself to do this constant questioning.

The training that helps us respond to change is sometimes internal, sometimes external. The mechanisms vary. Peer-to-peer dialogue, in its various forms, is often the best, and least exploited. ”

Norm Wold
Wold Marketing Group
Milwaukee, Wisconsin

“ The real problem is not if computers think, but if people do. ”

B.F. Skinner

“ Provide employees the opportunity for true participation in management decisions through focus groups. Employees are the true experts; they are now, or will be soon, performing the functions to make the new environment productive. Subscribe to the corollary of the *Peter Principle* that says, 'Decisions rise to the management level where the person making the decision is least qualified to do so.' Not to make use of all available points-of-view during periods of change will only lengthen the overall process required to make the change work. Change results are inevitable. If we do not plan the results, somebody else will. We might as well get the involved people planning from the beginning. People who are involved in the process tend to be committed to the results. ”

Tom Payne, writer, speaker
Lodestar
Albuquerque, New Mexico

VISUAL IMAGES

CONCEPTS AND DESIGNS

8

Once the parameters are established for a publishing project, the Client and the Creative Group collaborate to determine the creative direction. An original concept is developed, which is defined as:

- a concept or idea that is inherently different from other concepts or ideas.
- one that was not seen elsewhere by the creator (although all ideas are the combination of pre-existing elements put together in a new way).
- that which contains no parts of recognizable pre-existing copyrighted work.
- written or visual elements that convey a theme, direction, or campaign. It is represented initially by presentation materials, and ultimately by finished artwork.

The American Institute of Graphic Arts (AIGA) is a national graphic design association with local chapters. For information: AIGA, 164 5th Avenue, New York, New York 10010, 212/807-1990.

The American Center for Design is also a national graphic design association based in Chicago. For information: ACD, 233 East Ontario, Suite 500, Chicago, Illinois, 60611, 312/787-2018.

ORIGINATING CLIENT ROLE

To facilitate communication in the creative phase, the Client:

- provides all the background information necessary.
- understands concept ownerships (which should be in writing to avoid any potential misunderstanding).
  - If the Client originates a concept and hires a Creative Group to execute it, the concept remains the property of the Client.
  - If the Creative Group originates the concept, it is the property of the Creative Group (see Chapter 28, page 158).
  - If the concept is the result of collaboration between the Client and the Creative Group, it is assumed to be owned by the Creative Group, unless other agreements are made.

MAJOR DECISIONS

- plan project
- solicit proposals
- choose Creative Group
- finalize the proposal
- **select the design theme**
- **approve the design**

It is best to have someone from the Creative Group present at all presentations to decision-makers, even those that the Client Group makes to its upper management. The Creative Group may learn a great deal from people's reactions and comments to concepts that can be very useful in the rest of the work.

There are cases of a simultaneity of ideas—the same conclusions drawn from similar parameters. The one to publish first has the advantage. It is possible that two designers may conceive of the same design. For example, two different scientists independently invented the telephone, but Alexander Graham Bell was the first to get it patented and to the market. The most famous corporate identity case is the NBC logo—simultaneity between NBC and a Midwestern educational TV station.

Designers need an ever-expanding skill set to best serve their Clients. Companies hiring creative sources who utilize technology will want their publications to integrate with their entire marketing campaign. There will be a savings for them if the images developed can also be used for presentation, broadcast, and display purposes (see page 79 for multimedia).

The challenge for the graphic designer is to be conversant in these other media. In the past, design education was implemented during college and through apprenticeship. Then, the designer perfected skills through experience. Today, designers needs to know almost everything they needed to know in the past, but their education never ends.

- arranges presentation meetings with all decision-makers.
- ensures that someone from the Creative Group is present during presentations.
- approves all concept and design directions before work continues.

The Client is responsible for any appropriate legal searches, such as a trademark search or corporate identity search, for the concepts they select.

- This ensures originality and legal use.
- In the rare case of concept simultaneity
  - the Designer is not paid for creative fees, and must create another concept.
  - the Client reimburses the Designer for out-of-pocket expenses related to development of the first concept.

CREATIVE ROLE

After reviewing background information provided by the Client, the Creative Group

- contacts the Client if information provided is incomplete.
- creates original approaches to the assignment.
- presents concepts to the Client in the form of comprehensive presentations ("comps"—which may be sketches, mockups, or preliminary color representations) (see Glossary for further list).
  - The form and size of comps should be specified in the proposal submitted by the Creative Group.
  - Any concept, comp, or design that the Client does not choose remains the property of the Creative Group.
  - The Creative Group retains ownership of all comprehensive presentation materials.
  - All decision-makers from the Client Group need to be present during presentations.

- utilizes the approved design as a framework for production plans.
  - Concepts should fit within the budget range agreed to in advance.
  - Any additional costs for a given design are to be reviewed and approved by the Client before the work is performed.
- solicits suggestions on the most efficient production methods for the approved design from others in the production cycle, which will vary from project to project but usually include:
  - Writer
  - Photographer
  - Illustrator
  - Page Composer
  - Production Coordinator
  - Printer.

As a design tool, the computer allows a designer much more flexibility than conventional means. Designers can make the best use of the technology by going to the workstation with a concept already in mind. Then they can try variations, but need to be careful not to spend too much time on "what ifs," as the design budget can be used up very quickly this way. Sometimes conventional means may be faster than the computer for certain functions.

PRINT ROLE

It is important for the Print Group to see the comprehensive presentations to give input for production procedures. Because the Print Group has already provided estimates, they may

- advise on the best way to prepare artwork.
- recommend modifications that may make the design more efficient to produce.
- need to revise their estimates and prepare final quotations according to the approved design.
- meet with the Creative Group to begin the production planning process (see Chapter 17, pages 95 and 96).

VIEWPOINTS:

CREATIVITY AND SURVIVAL

"To communicate effectively and build a distinctive identity, a corporation must develop its own "voice," a voice that is instantly recognizable; a voice that conjures up strong images and associations; a voice that establishes the tone of the company's communications: classic or innovative, glamorous or functional, aristocratic or irreverent, elegant or just simple and down-to-earth; a voice that fuses language, design, and content to convey the company's distinctive personality and culture."

Ronald R. Manzke
Managing Director
Siegel & Gale/Los Angeles
Los Angeles, California

"If a committee has a strong leader, the members of the committee can inform the leader and it can work well, particularly if the leader is in an area the leader is not familiar with and members of the committee are. That is where a lot of designers are lacking in skill: managing a presentation and organizing material so that you can manipulate their attention. You are on a stage and just as an actor will manipulate his audience to evoke responses, you are doing the same thing. You can't simply let your work speak for itself because it won't. You have to set the stage and that is an art. A designer has to be good at both coming up with the ideas and also presenting them—and most people are good at only one or the other. There is always a danger of someone who didn't actually do the work botching a presentation. Also, you have to get access to the one who is making the decisions. If not, good luck."

Kevin Shanley, designer
SWA Group
Houston, Texas

"Clients purchase equipment, and, in doing so, become a graphic artist's competitor. Artists buy more equipment and thus expand the ripple effect of competition.

I've learned that we full-service design firms need to reposition ourselves and our businesses to compete as a viable business in the future. We need to do a better job setting fees and accounting for our costs.

The future of design is in creative design; it is in problem-solving; it is in being professional enough to know what tools and resources are appropriate for a given challenge, and arriving at effective solutions for our clients."

Daniel Dejan, designer
Dejan & Associates
Chicago, Illinois

"It [the computer] does some things easier and faster than by hand, but it takes time to learn and to solve problems that come up. I feel it can be detrimental to creativity because of the tendency to become so involved with type and kerning and details—the overall design is not focused on as clearly. However, the speed makes it possible to try different ideas, print them out, and see how they look.

Because the laser-printed page is so much more finished looking than pencil or marker layouts it provides a much clearer idea of the end result, but also calls attention to bad spacing and other details which shouldn't be a concern in the initial stages. The designer must resist the temption to fiddle with it at this point and remain focused on the concept."

Carole Taylor, designer
Carole Taylor Design
San Francisco, California

"It's our skill as designers, not as computer users, that will set us apart. We will go further concentrating on the craft of design and less on the tools. Quality is ultimately tied to design. The computer only allows you to express the talents you already have, and gives you speed and control in executing projects."

Robert Vann, designer
Lake Mary, Florida

“The skills of the designer are devalued by the software developers. As designers, we should d be thinking about how much space we should delegate to each element, how elements should interact, what typeface should be used, what is the best way to relate the message to the viewer—not what 'skuzzy' port my Syquest drive is plugged into, or what file format I need to create color separations out of FileMaker Pro.

Do we complain about technology? No. We revel in it. We find that we have what we've always been missing: control. We control a project from beginning to end. We can produce design in record time. What is the price for this control? Losing design time due to learning software, debugging software, and updating software. More time is spent learning to typeset, scan photos, color correct, composite, and trap than on design. On the up side, the more we learn to do well, the more we can do in-house and bill for it. On the down side, the more we do in a shorter time, the more is expected.”

Eric White, designer
Interactive Communication Arts, Inc.
Winter Springs, Florida

“The wonderful thing about traditional design is that it is slow—it gives you time to think. The bad thing about desktop publishing is that you can do bad design so quickly.

Amateur designers don't trust themselves to keep it simple. They believe they can prevent a piece from looking boring by 'dressing it up.' So you have a combination of utterly standardized two- or three-column structures filled with an unpredictable variety of oddments tossed together like fruit salad, trying desperately to be original and innovative and—heaven help us—*creative*. It would be far better if they'd forget about being different and concentrate on making the message understandable. Go for clarity, not beauty. We live in an era when all the concentration is on the medium. But the medium is not the message. The message is the message.”

Jan White
designer, author, consultant
Westport, Connecticut

“ Most designers will be disappointed when I reject a particular direction, but the real professionals realize that I don't do it arbitrarily. There are solid reasons why something won't work. I don't just say, 'We have to change this.' Instead, I'll say, 'It would be better if we emphasized this aspect because, for these reasons, it's more important than what you've shown.' Ego is always involved in design. But when you can separate your ego from your design—when you don't take it personally, and you realize there's a reason for making changes—then I think you become a much better designer, certainly a better business person. ”

Kathleen E. Zann, Manager
Marketing Communication
James River Corporation
Richmond, Virginia

“ There is a myth that there's big money to be made in desktop publishing. If you believe this, you'll believe anything. The people that control the money think that because the computer can do any publication easily and quickly, there's no reason to pay anything more than the minimum for a computer operator. Consequently, the salaries for desktop publishing and computer production tend to be low due to the number of people vying for those jobs. Skilled designers' salaries also seem to be stagnant. Freelancers are finding that much of the work that was formerly farmed out is now done in-house, often by someone who's been upgraded from a secretarial position.

It appears that to become a designer or desktop publisher means doing it for it's own sake, not to make a killing. If you want to earn a decent living, management is the place to be, but you won't be doing much of the desktop publishing or design in that capacity. ”

David Doty, designer and author
PageWorks
Chicago, Illinois

“ We should no longer attempt to compete for projects that can be produced more economically by the client in-house, such as one- or two-color newsletters or flyers. Instead, we need to concentrate on the kinds of projects that require strong graphics and strategic approaches.

Rachel Deutsch, designer
Elliott Van Deutsch
Washington DC

"We should all become a little humble with the tools of computer technology, but take solace in the realization that design principles stay steady and that the artistic foundation is secure. Creativity is still the domain of our personal human choice. Our minds and hands still rule the future of design; it's the wild stallion of technology whose reins we have to keep learning how to hold onto."

Gregory F. Golem, designer
United Services Life Insurance
Arlington, Virginia

"The three lessons I learned about design at Apple were: (1) design is not about creating a thing; it is a process; (2) the Macintosh is more than a tool; it is a medium; and (3) there are no new ideas; technical innovations are nothing more than a series of ideas linked together by association—much like the design process. Technology has always changed the way we experience the world. And, therefore, the way we design and create. And, ultimately, the things we design and create. And the way we work with one another. If the computer is the great equalizer, where do we as designers add value? First, we cannot be specialists anymore. We need to understand the macroview of the client's needs. Perhaps the best combination is to be a generalist as to types of projects but be industry specific so that clients do not have to pay for a steep learning curve. Second, be ready to take on one-of-a-kind projects or systems projects. Third, work more collaboratively internally and externally, with projects being more process driven and less 'free form,' and get used to working with a broader range of people (programmers, animators, interface designers) than in the past. Finally, don't just focus on computing technology as the only area that will alter the way we design. Interesting and exciting things are happening with broadcasting and telecommunications technology. The changes there will have enormous impact on what we do and where we can add value."

Clement Mok, designer
Clement Mok Designs
San Francisco, California

“ A recent notice in a magazine asked for desktop publishers to fax their resumes to the publisher of the magazine. Whatever happened to portfolios? Visual content? Designers understand the importance of computers and programs, but it isn't so certain that techies and some clients understand the importance of design. ”

Tad Crawford
Publisher and Writer
Allworth Press
New York, NY

“ Shaping perceptions is the heart and soul of design. Color, image, type, paper texture, shape, and size work together to convey one or more messages, which can reinforce, supplement or even replace text. Designers are becoming increasingly sophisticated in using perceptual tools—by choice and by necessity. They recognize the growing importance of perception in communicating with today's global audiences, with their shared and divergent values and changing ways of processing information. ”

Frankfurt Gips Balkind, NY/LA/DC, Simpson Paper Company
Fairfield, California

“ I provide designers with all the technical background they need. If they're not out in the cold, if the purpose of the project and its audience are thoroughly understood, they should be able to solve the problem. I don't want to go through a lot of iterations and be presented with tons of sketches or comps. If there's enough verbal interaction and agreement before we actually execute something, there's no need for them. All I need to see is one design concept or maybe two. ”

John J. Dietsch, Director
Corporate Communications
Booz-Allen & Hamilton Inc.
Bethesda, Maryland

“ We want to see eight to ten logo ideas. It would be unacceptable if someone came in and announced, 'Here it is; here's your logo.' We need to see options, but they don't necessarily have to be eight to ten completely different solutions. A few variations enable us to say, 'Yeah, we like this one,' or 'Maybe you need to go back and refine these two.' Sometimes if we like several designs, we might ask the designer to take a particular aspect from one and combine it with another. ”

John and Bill Schwartz, owners of Schwartz Brothers Restaurants, Washington state

Lynn Finch, President
CADTEK Services, Inc.
Maitland, Florida

“ Skills a designer needs beyond design: communication skills; organizational abilities; business savvy; team work; psychology behind the image; computer knowledge; partnering with suppliers. ”

Nancy Miller, designer
Studio One Graphics
Irving, Texas

“ Today, the success of a graphic designer depends upon creative ability and expertise in the hands-on basics of production and design, combined with the skillful use of computer technology to explore new artistic avenues. There is a real need for a graphic designer to be constantly aware of any changes in technology, not only in the design field, but also in the printing/production processes.

In order to maintain a competitive edge, advertising agencies and design firms now demand computer design skills as a prerequisite to employment, as opposed to the past practice of allowing a designer to learn those skills while on the job. There is a great amount of pressure on a designer to implement the best technology to enhance production speed without sacrificing design integrity. ”

Alyce Kaprow
computer graphics consultant
Newton, Massachusetts

“ We are challenged to somehow seamlessly continue the excellent creative work we do, learn a new set of skills to integrate the computer into new areas of design media, and master the expertise in production tasks once reserved for prepress professionals. Though technology may offer tools to do much of the work, it is forever linked to the human controlling it. Should one person be responsible for the skills needed to control such a complex sequence?

With the computer, the designer is more likely to carry the creative steps into the mechanical/production phase directly. This creates an atmosphere where no one person is relegated with the responsibility of doing only mechanicals, and no individual can strut one's ego claiming to be superior to such lowly tasks. Managing jobs and egos takes on new meaning in this age of technology. ”

“ We need an environment where the quality of design does not get lost in the rush to complete the project; where artists are not forced to comply with the demands of technology. It is up to the people involved to develop a workflow that promotes a synergy between deadlines and the design process. One method that works well is to team up designers with technologies that will produce art, not desktop punishment. ”

Eric White, , designer
Interactive Communication Arts, Inc.
Winter Springs, Florida

“ I like designers who beat me up. I don't respect people whose response, when I say, 'Let's do it this way' is 'That sounds good.' That's not the point of the whole process. If I knew what I were doing, I wouldn't have to ask designers to come in and help, and be willing to pay good money for it. But when you're beating me up, I want to know why. I want to know why my idea is really not good. I want to know what your idea is going to contribute. What's really important is how much designers bring to the party—the totality of their contribution. The best projects that I've done involved designers who have really contributed a lot, I mean, a lot beyond the look of the thing. And the more the designer brings to the party, the more likely it is that we're going to party again. ”

Robert Moulthrop
Vice President
Scudder, Stevens, & Clark, New York formerly market director, KPGM Peat Marwick, New York

“ There are a couple of reasons that top management's concern and participation ought to be mandatory. Communication design (like design in general) is crucial to a corporation's success, yet not likely to be understood throughout the corporation. Until it is understood, it needs support from the top. And because it depends on a combination of broad concepts and extremely fine details in execution, top management must be mindful of the vigilance required to keep a good program from eroding. Top involvement is the irreducible minimum requirement for long-range communication effectiveness. I can think of no exception to the rule that where there is a sustained communication program of distinction, top management is directly involved. ”

Ralph Caplan
communications consultant and writer, New York City

9

COMMISSIONED ILLUSTRATION AND PHOTOGRAPHY

The authorization to create visual images, commissioned specifically for a publication, is usually given to the Art Director by the Client. The Art Director then chooses the Illustrator or Photographer.

MAJOR DECISIONS
plan project
solicit proposals
choose Creative Group
finalize the proposal
select the design theme
approve the design
review the writing outline
edit the first draft manuscript
review the revised manuscript
review the final manuscript
approve visual components

ORIGINATING CLIENT ROLE

As part of the design process, the Art Director presents the concepts for visual images to the Client. The Client then:

- selects a design or concept for the visual images.
- approves the visual style and the selection of illustrator or photographer.
- reviews the budget with the Art Director for any change from the estimate.
- reviews usage rights with the Art Director.
    - The Client purchases the use of images only for the purpose specified in the proposal.
    - If unlimited uses are needed, arrangements are made (see Chapter 28, page 159).
    - Agreements should be defined before the work is done.
- pays any cancellation fees if work is stopped in progress (see pages 50 and 51).
- is available for approvals in progress, especially for photo sessions.

This is a very important point at which to stop and check the design against the budget. The proposal or initial estimate can only approximate costs, for the estimate is made before the design concept exists. Most proposals should contain a not-to-exceed figure, but occasionally the best concept may require additional resources, such as more involved photography, illustration, manipulation, etc. The Creative Group advises the Client of this during the presentation. It is up to the Client to decide if they wish to spend the extra money for the execution of a considered concept.

CREATIVE ROLE

The photographer, the illustrator, or both, join the Creative Group under the direction of the Art Director. They may be subcontractors or on staff.

- The Art Director commissions the creation of visual resources, and
    - secures Client approval on concepts.

- shows designs and specifications to the Illustrator or Photographer.
- specifies to the Illustrator or Photographer all financial agreements and arrangements in advance, including payment terms.
- receives and reviews all preliminary work, such as pencil sketches or Polaroids®.
- shows this intermediate work to the Client for approval.
- supervises the direction of finished artwork or photographs (which often includes on-site supervision during photography sessions).
- provides approvals if the Client is not available, especially under deadline conditions. The Client may agree to delegate the decision to the Art Director at the start of the project. (If this is the case, the Client accepts the financial implications of the Art Director's decisions.)
- manages the process and receives final materials. It is easy to lose track of original images during the production process. Therefore, any transfer of images from one place to another should be documented in writing by the Creative Group.

The Graphic Artists Guild is an association of illustrators, photographers, and graphic designers, with local chapters. For information: Graphic Artists Guild, National Office, 11 West 20th Street, 8th Floor, New York, New York, 10011, 212/463-7730.

- The Illustrator works according to the predefined specifications provided by the Art Director and
  - provides rough pencil sketches to the Art Director prior to creating finished artwork.
  - creates finished artwork under direction of the Art Director.
  - has a cancellation fee arranged in advance. Usually, if an illustration is cancelled after work is in progress (for reasons other than the Illustrator not working within the parameters), the Client pays a 50% cancellation fee plus expenses.

The Advertising Photographers of America (APA) is a national association of professional photographers. For information: APA, National Office, 7201 Melrose Avenue, Los Angeles, California 90046, 213/935-7283.

- The Photographer works to the predefined specifications provided by the Art Director, and
  - often provides Polaroids® (created during a site visit, if on location) to Art Director for approval prior to the photo shoot.
  - creates finished photographs to specifications provided by the Art Director, who usually is on site providing direction. It is advisable that the Client be present during the initial part of photographic sessions, in case content decisions need to be made.
  - has a cancellation fee arranged in advance. Generally, if a photographic session is cancelled by the Client more than 24 hours before the scheduled time, the Client will not be charged, except for large guaranteed bookings involving travel. If the Client cancels the photographic session less than 24 hours before the scheduled time, the Client will be charged a 50% cancellation fee and incurred expenses up to that point. This is because the photographer may be unable to book any other projects for that time slot. (This is according to general photographer's trade practices. See the individual photographer's estimate form, as these practices may vary).

PRINT ROLE

Often, the Print Group can give advice on the best production process for a given approach or effect that the design dictates. This can add efficiency and cost savings later. The Art Director facilitates this by showing designs to the Print Group in advance of creating finished artwork or photographs.

VIEWPOINTS:

ORIGINAL IMAGES

"Twenty years ago, no one except fashion magazines took art directors seriously. Now, the role of design in publications has become crucial. Most magazines put art directors at the top of the masthead. Many publications still make the mistake of assuming that 'content' refers strictly to what's written. Visual content is critical now, and print media's future depends on a merging of the written and visual. We haven't seen many major magazines edited by former art directors, but I think that will change soon.

Baby-boomers grew up with television, and that influenced publications. But even TV has trouble grabbing our attention now. It's paradoxical, but the future of print actually depends on creating more visual content."

Roger Black, designer
Roger Black, Inc.
New York City

"When I lobbied on behalf of the Graphic Artists Guild for the copyrightability of typefaces, the Authors Guild opposed us because they feared that authors would find their ability to express themselves in words to be limited if typefaces were protected. Unrealistic as this argument was, I think that it may have foreshadowed what is now a noticeable phenomena: the increasing content of design. Regardless of the degree to which typefaces and designs can be protected by copyright, it is certain that the impact of the visual content more and more works in synergy with the impact of the written content."

Tad Crawford
Publisher and Writer
Allworth Press
New York, NY

10 STOCK IMAGES: PHOTOGRAPHY AND CLIP-ART

"Stock" refers to images that already exist, either as outtakes from previous creative projects, or made specifically for sale as stock. Stock images may exist in traditional or digital form, and many are available through a stock agency. Key issues are ownership and permissions for use.

- The creator may own the copyright to the work.
- A stock agency may own:
  - the copyright if agreed to by the creator.
  - the physical software or image, but not the copyright if the work is in the public domain (see next page).
- The owner may sell the copyright or may grant permission to use the image under specified conditions. Fees will vary by terms of the agreement, which may specify
  - intended use (relative prominence, as on a brochure cover, size of image, etc.)
  - intended audience
  - quantity of distribution
  - time period for use
  - degree of exclusivity (see Chapter 28, page 158 and Chapter 29, page 163–164).

The need for permission to use an image depends on use.

- Permission is needed when:
  - the image is used as reference by another artist in developing a new image. This is true even though the resulting image may not appear to be derived from the original.
  - the image is included in presentation materials, such as comprehensives to show the Client. The owner generally does not charge a fee for permission to use an image in a presentation. The exception is that some stock agencies may request a modest presentation fee. If new images are to be created based on the image used for a presentation, the creator of the presented image should be hired to execute the new images. (If the creator is not available or unaffordable, the concept should be changed).

Many Designers photocopy or scan images from a magazine or stock catalog to include in their comprehensive presentation materials when showing a design to a Client. The Designer should ask the owner for permission. Generally if used for presentation with intent to obtain an image from the owner if the concept is chosen, then the need for permission is lenient. Also, permission may be implied by the format of the source material. For example, low-resolution stock photography images on a photo CD are *intended* for use in presentation materials. Once an image is selected for use, the stock agency provides a high-resolution image and charges a fee for use. When in doubt about when to obtain permission for presentation, the Designer should always contact the image's owner or agent.

- the image is scanned into a computer.
- the image is reproduced in whole or in part (see Chapter 29, page 164) for any purpose including printing or presentation.

- Permission is not needed when:
  - the image is to be maintained in Creative Group files for future reference, such as for inspiration or future consideration.
  - the image is in the public-domain (see Chapter 29, page 162), and is reproduced directly from an old book or magazine. Generally, images published more than 75 years ago have lapsed from copyright protection. However, this is not always true. And older image may be protected as part of an estate.
  - the image is in the public domain, but a stock agency owns the tangible artwork. Once purchased from a stock agency, it may be used any number of times without further payment.

ORIGINATING CLIENT ROLE

To illuminate publications, the Client may wish to utilize pre-existing images.

- The Client may choose to locate public-domain images and verify legal usability.
- The Client may ask the Creative Group to obtain images. The Creative Group may charge the Client for search and purchase fees.

CREATIVE ROLE

If directed by the Client to locate existing images, the Creative Group most often deals with stock agencies.

- The Creative Group works directly with the stock agency regarding
  - the search and selection of images
  - the negotiation of uses and fees.
  - the transference of original or high resolution images.

MAJOR DECISIONS

- plan project
- solicit proposals
- choose Creative Group
- finalize the proposal
- select the design theme
- approve the design
- review the writing outline
- edit the first draft manuscript
- review the revised manuscript
- review the final manuscript
- **approve visual components**

- The Creative Group is responsible for the
  - management and care of original materials while they have possession or use of them. It is easy to lose track of original slides or artwork. Therefore, any transfer of images from one place to another should be documented in writing by the Creative Group. Digital images are not as perishable because a copy can be as good as the original. It is imperative for the stock agency to have a good backup system.
  - return of all original material according to the agencies' practices.
  - financial reimbursement to the stock agency if the image is damaged or lost (see Chapter 27, page 153).
- The Creative Group communicates the following to the Client:
  - costs for the entire stock package or packages, such as collections, books, or disks from which any image is used. Often this will require reimbursement from the Client.
  - the difference in cost between stock versus commissioned images—generally stock is less expensive.
- The stock agency should specify to the Creative Group whether they are providing an original or duplicate from an original of a digital file. It should be a duplicate.
- Many collections may now be ordered on CD-ROM. The Creative Group may choose what images they want to unlock the rights to use for finish with the stock agency.

PRINT ROLE

The Print Group is responsible for maintaining any images received from the Client or Creative Group in good condition.

- The Print Group is responsible financially to the Client or Creative Group, whichever has contracted with the stock agency, if an image is damaged or lost, unless by situations for which the Print Group has no control (see Chapter 27, page 153).
- The Print Group must return all original material to the Buyer at the conclusion of the project.

VIEWPOINTS:

REPURPOSED IMAGES

“Kodak is taking an active role in protecting a photographer's stock images. We are embedding an encryption code that will lock images at low resolution. It is like watermarking an image so they can be viewed but not sent.”

Dave Hazlett
Marketing Education Center
Eastman Kodak Company
Rochester, New York

“The American Society of Media Photographers has taken an active role to protect the rights of photographers in stock images. By creating the Media Photographers' Copyright Agency, ASMP has taken an important step toward raising standards in this age of digital reproduction and manipulation.”

Tad Crawford
Publisher and Writer
Allworth Press
New York, NY

11 LIBRARIES AND DIGITAL COMMUNICATIONS

Unfortunately, what holds back the promise of utilizing the potential of libraries today includes:

> expense—both in converting non-digital images to digital and in the hardware capabilities.
> hardware wiring and cellular capacities.
> format and software standards—there are many competing standards being proposed by a variety of organizations.
> government regulations for telecommunications—including noncompete controls and rate structures.
> competing industries—industries that were once distinct (telephones, computers, broadcast television, cable television, newspapers, entertainment, information) are now overlapping and competing for similar markets.
> compression of data—image file sizes and transfer through hardware limitations.
> speed of transfer methods—cellular is slow compared to cable but cable installations are limited.

The world is becoming more visual. Electronic documents are becoming more common, and the methods for storing, defining, and sorting continues to grow. Digital technology allows documents to become part of text and image databases to be cataloged, referred to, updated, altered, used in new ways, analyzed, and incorporated into bigger communications systems. When the Library of Congress accepts only publications available in both electronic and hard-copy forms, this vast resource of information will be accessible to everyone tied into computer networks. Ours is the generation to translate publications from analog to digital, and the technology will continually accelerate the delivery of information to the technologically astute community of users. Many forms of libraries will join in this escalation:

- public libraries on bulletin boards with public domain images. (Most bulletin boards are currently text-only, but capabilities will grow with developments.)
- electronic subscription services, distributed on-line or via CD-ROM.
- text, information, and image databank companies that search requested subjects.
- private network libraries through servers and used in a department or company-wide which can be:
  - format specific.
  - locked to control accessibility per appropriate user (Chapter 14 page 74].
- personal libraries for specific workers who develop their own images and then categorize them for later use.

MAJOR DECISIONS
plan project
solicit proposals
choose Creative Group
finalize the proposal
select the design theme
approve the design
review the writing outline
edit the first draft manuscript
review the revised manuscript
review the final manuscript
approve visual components

ORIGINATING CLIENT GROUP

Maximizing business publications through a digital library enables a sharing of information and resources. Digital networks makes it to be easy to find, sort, update, store, transmit, and manage documents—but only through effective management and library software. There are many ways to set up such a document library:

- maintain a text-only library, indexing originals.
- have a software-independent management system that allows whole documents to be viewed in their original

format, but not to be alterable.

- provide a library network complete with lock-outs and security codes to determine depth of access.
- provide a data base of training and resource information to assist workers.
- utilize a bulletin board and storing system to communicate time-sensitive information.
- convert some documents to electronic distribution, such as directories, schedules, and employee lists—documents that tend to change quickly.
- commit information to electronic format which takes up less storage space than paper files.

Libraries offer many new ways of working and managing information. To utilize them most effectively requires a cultural change in how users relate to documents. For example, some information never needs to get printed, but for the audience to accept purely electronic information means a different level of receptiveness.

Accessing non-original created information, text, or images requires a sense of responsibility. To avoid digital liability (see Chapter 28, page 156 on ownership of materials):
> do not scan images into the computer and use them in recognizable ways.
> do receive permission from the copyright holder if you do want to use an image (see Chapter 29, page 162).
> do inquire about the source of employee- or supplier-provided images.
> do protect your own proprietary images through policies of use and employee training.

CREATIVE GROUP

It used to be that an artists' palette was filled with globs of various colors that could be mixed into a world of hues. Today, the artists' palette is filled with the world of images that can be combined in infinite ways: possibility is only held back by imagination. Such libraries are built through:

- creating original images that can be used over and over in various applications.
- subscription series received via CD-ROM.
- on-line accessibility to libraries.
- stock image searches.
- cataloging of created publications for reference.
- scanning found images (must be in the public-domain—see Chapter 29, page 162).

One challenge to having an effective digital library is to be able to find the images that you want to use when you want to use them. Many software products display miniatures of the documents, have key description words for sorting, and have sophisticated management and storing capabilities.

PRINT GROUP

Libraries and digital communications may push the printing industry into a new role. It is not that digital communications will *replace* print, but the way print is used will evolve. Providers of print services can expand to offer a range of new capabilities to their clients, thus remaining viable as the technology affects print volume and use. New services can include:

- storing and retrieving various documents that were printed, and cataloging them through a library software.
- linking into digital printing.
- providing high resolution replacement service for low resolution library images.
- providing resident data base services of images to clients.

Responsible Usage:

When utilizing images within documents, files, presentations, or publications, responsible usage means not only working responsibly, but also ethically and without risk. Here are guidelines to help when using images:

> use original material whenever possible.
> when using a found image, track down the source and make contact. If you can't negotiate with the originator, don't use it.
> negotiate, negotiate—everything is subject to situation.
> get permission in writing—put in your correspondence: 'Unless I hear from you otherwise . . .'
> don't assume you can change a found image beyond recognition unless you are *completely* recreating
> when in doubt, get permission.
> remember permission fees cost less than legal fees—don't hope you won't get caught
> don't use celebrities—many older images, such as Einstein and Micky Mouse, are *not* in the public domain and are aggressively watch-dogged by their owners.
> be clear on fair use—if you can make money by using a found image, have a lawyer check out ownership or do the homework yourself.
> with fine art of any age, get permission from the owner, whether that is a museum or an individual.
> if you want to use a government-produced image, check with the National Archives, Library of Congress, or the White House Press Office for permission. There should be no fee.

VIEWPOINTS:

PALLET OF IMAGES

“The driving force behind development of image databases is the rapid growth of digital photography and video. Computer-based photography, in particular, is expected to skyrocket with the widespread availability of Eastman Kodak Co.'s Photo CD system, which enables anyone to take a roll of 35 mm film in for processing and get back a CD-ROM containing high-resolution digitized photographs.”

Aileen Abernathy
MacWeek
San Francisco, CA

“Indexing or keywording is the most important aspect of an on-line library: how you get to it.”

Sandra Kinsler
PhotoLibrary Management Service
Ventura, California

“Ours is a ten-person networked shop. We're concerned with our clip art files: keeping versions straight and up-to-date, controlling what happens to the file, and getting the file back where it belongs. We have an employee who spends 50% of her time monitoring what is going on, identifying pot holes, and coming up with solutions. We try to find out how to make these files accessible, test new versions of software, and to understand the features. She makes mini-training materials for others so we can cut down on training time.”

Scott Hazel, design manager
System Solutions
Washington DC

“In a network environment, a server-resident image database minimizes disk-clogging duplicate files, ensures that everyone is using the same version of a graphic, and eliminates the frustration of finding that another user has borrowed a needed graphic.”

Aileen Abernathy
MacWeek
San Francisco, CA

“Fetch's [by Aldus] multi-user capabilities boost productivity because several people can work on images simultaneously. For a big project that has to be done very quickly, one person can finish the details of a graphic while another proofreads its text. Fetch supports multi-user access.”

C. David Piña
Piña Design
Burbank, CA

DEVELOPMENT

12 WRITING

Conceptual direction and writing may either begin the project or follow the design concept, enabling an integration of verbal and visual messages. Making sure that the writing and design are integrated is best done as a collaborative process between the Client and the Creative Group. It is usually prepared in stages:

Independent Writers of Chicago (IWOC) is a good source for writers to interact with their colleagues. For information: IWOC, 7855 Gross Point Road, Skokie, Illinois, 60077, 708/676-3784.

- Research, which includes reading previous publications, investigating sources for information, and interviews.
- Outline development, which may give suggested page breaks.
- First draft manuscript for the Client to read and edit.
- Revised manuscript, which incorporates Client's edits or Editor's edits.
- Final manuscript preparation, which ends the writing process.

MAJOR DECISIONS

- plan project
- solicit proposals
- choose Creative Group
- finalize the proposal
- select the design theme
- approve the design
- **review the writing outline**
- **edit the first draft manuscript**
- **review the revised manuscript**
- **review the final manuscript**

ORIGINATING CLIENT ROLE

The Client both initiates the writing and serves as a resource for its development.

- The writing may be commissioned in one of three ways:
  - the Client contracts directly with the Writer
  - the Creative Group provides the writing
  - the Client handles the writing. This often happens when content is of a technical nature.
- To initiate the research for writing, the Client
  - assembles content information in verbal, handwritten, typewritten, digital, or printed form
  - assembles supporting materials which may include the following:
    - company information
    - articles on the industry
    - competitor's literature
    - market research data
    - slides and videos
    - other appropriate information.

- The Project Manager is responsible for setting up any needed interviews with company employees and with customers, where appropriate.

No matter who generates the writing, the Writer, the Editor, or both need to work closely with the Client, Art Director, and Designer from the beginning to ensure the integration of writing with the design.

It is the responsibility of the Client to:

- proofread the final manuscript (check the text for any errors).
- obtain all approvals by decision-makers before the design proceeds. This yields the most efficient and cost-effective result.

One of the greatest potentials for extra costs that Clients incur happens when they prematurely turn a manuscript over to the Creative Group before they have all needed approvals. This is often tempting to do in the interest of getting a "head start." But time is generally lost later through alterations. It is worth taking the time at the beginning to have all decision-makers "sign-off" on the manuscript. If they understand that changes will add significantly to the project costs, decision-makers will be careful to approve the writing at an early point. This also ensures that the decision-makers are part of the process and will better anticipate the next phases. It's easy to think that typographical alterations are easy and cheap with electronic publishing, but it's the *ramifications* of changes that can affect the project in major ways—from placement of design elements, to the number of pages.

CREATIVE ROLE

The Writer is part of the Creative Group, whether

- in-house with the Originating Client
- inside the Design Firm
- an independent contractor.

When the Client makes the assignment, it is best to review with the Art Director and the Designer each of the following stages in the writing process.

- The Writer begins by doing appropriate research:
  - reading past company materials.
  - gathering information.
  - collecting factual data.
  - working with the Project Manager to set up interviews and coordinate meetings.
  - interviewing appropriate people such as Client management staff, customers, and suppliers.
- Next, the Writer submits an outline
  - to the Creative Group first to see how it complements the design theme, or
  - directly to the Client and then to the Creative Group.

Six pairs of eyes should look at each project before it prints or is finished:
- writer—who generates the information initially
- editor—who reviews the content, style, and grammar
- novice—someone who knows little about the subject who can provide a fresh viewpoint and understanding
- desktop publisher—who places the copy into pages
- proofer—who reads for correctness
- editor—who checks over content and correctness

It then usually goes back to the desktop publisher again, through the proofer, then the editor, and back again if needed. Most projects go through the last three professionals a varying number of times.

- Once the Client approves the outline, the Writer creates the first draft manuscript. The Client and Creative Group review this against the design and request any needed revisions.
- The Writer prepares a revised manuscript that incorporates all changes from the Client and Creative Group, who review it again.
- The Writer uses feedback from the revised manuscript and prepares the final manuscript.
- The Editor manages all phases of the editing process and checks for:
  - typographical errors
  - grammatical correctness
  - stylistic consistency.
- Additional drafts or research may be requested by the Client and require extra cost. As always, the Client must be apprised of additional costs before work begins.
- When the Client approves the final manuscript, the design phase begins (If done before final, extra costs may be incurred for alterations.).
  - The Writer provides the Creative Group with a final proofread version on computer disk with matching a hard copy.
  - The Creative Group transforms the manuscript into the page design and reviews it for conversion errors. The Creative Group is not responsible for typographical errors.
  - Ownership rights are similar to the design rights (see Chapter 28, page 158).

PRINT ROLE

The Print Group is generally not involved in this phase.

VIEWPOINTS:

VISUAL LITERACY

“Editorial judgment remains a high-level task. But if document design is ‘cookie cutter,’ as it is in our manuals, then there are few design decisions and fewer personnel required.”

Helen Herber
Associated Systems Incorporated
Chicago, Illinois

“We have found that it is difficult to train a secretary to do graphic design. They can be taught the computer, but the graphics capability is not there. Our new employees now have degrees in graphic design—it is a fallacy that anyone can do incredible work on the computer. It is just a tool.”

Pat McNamara
A.T. Kearney & Co.
Chicago, Illinois

“Writers should be concerned with visual literacy so they can communicate more efficiently and effectively: not only among themselves, but with designers and others as well. They will also begin to understand the value of visual literacy and the effort it takes to achieve it.

Desktop publishing can’t help but increase visual awareness. Whether it produces literacy is debatable. But it does make writers and others aware of what we do as designers.

Most designers I know are not concerned about writers learning to do layouts and design on the computer. It really isn’t any more of a threat than our having word processors makes us writers or poets.”

Hayward Blake, designer
Hayward Blake & Company
Assistant Professor
Northwestern University
Evanston, Illinois

“ In many cases, desktop publishing has put design in the hands of writers whether they want it or not, particularly for presentations and in-house documentation. At the same time, the threshold of expectation has been raised. Five years ago, it was sufficient to print a proposal on a laser printer because so many people were still using dot-matrix printers. Now, everyone uses laser printers.

Writers need to understand that their words are only one dimension of the message. Since they create the message, they should be highly motivated to learn how to use type, paper, illustrations, photographs, color, etc. to enhance rather than impede their message.

Although a lot of designers disparage the 'untrained' who are using desktop publishing, in the long run the technology will make the visually illiterate more aware. When the 35mm camera was made available to the masses, there was a great outcry by those who believed that flagrant use of the camera would destroy photography as an art. ”

C.J. Metschke, writer and trainer
Monterey Press, Inc.
Vero Beach, Florida

“ A computer has never created 'design.' It did not go to design school. It does not understand the human psyche. And it cannot expound on the virtues (or inconsistencies) of Nietzsche. The computer is a modern design tool, but to use it well you must be a designer adept at not only placement and aesthetic, but the idiosyncrasies of the machine. ”

Regina Rubino, designer
Louey/Rubino Design Group
Santa Monica, California

“ All clients are trying to stretch their communications budgets. Some wisely, by searching out design firms that provide value—top design at reasonable costs. Because of this, design firms that are strategic and conceptually creative, and have a strong grasp on technology's role as a designer's tool will ultimately survive.

Layout and typesetting software has become more readily available and accepted by untrained designers and non-designers. However, merely having desktop capabilities does not give a person the power to communicate a message effectively. To maintain the integrity of our industry, we need to educate clients about the role computer technology plays in design. ”

Nanette Wright, designer
Wright Communications
New York City

“ Desktop publishing does not stop visual illiterates from doing terrible things on paper—it can make it easier for them. What is wrong with computers is *not* the computers, it is the people using them. ”

John Rosberg
Rockwell International Corp.
Chicago, Illinois

“ Computer-aided design and production have enabled us to streamline the production process, but this means that increased project management skills are in demand. And keeping up with the technology is costly. We also see an information gap in designers' skills. Experienced designers are not fluent in the computer, and the learning curve is painfully long. On the other hand, computer designers first entering the profession are not skilled in client needs, business savvy, or managing their time and budgets.

There are opportunities emerging for designers to become strategic partners with their clients, providing sophisticated visual communication skills to guide the mass of in-house desktop publishing efforts. 'Good is the worst enemy of great,' and responsible graphic design doesn't stop at good. ”

Miranda Moss, designer
Yamamoto/Moss
Minneapolis, Minnesota

“ The PC has made a huge difference in how enjoyable it is to write. The ability to manipulate text the way you would clay in sculpturing is vastly encouraging to a writer. It's one of my great puzzlements that the quality of writing in the world hasn't improved in line with the opportunity computers have given for people to improve. ”

Glenn Rifkin interview with Teller, magician and comedian with partner Penn Jillette, *ComputerWorld* newspaper Framington, Massachusetts

“ Now our working milieu is in the world of 'RoboDesign.' Everything is based on speed: faxes, Fed Ex, modems, and the computer. Client design expectations have become marketing driven and more sophisticated, and they have a working knowledge of computer capabilities. Because of this, we still use the basic process of design, yet we get there much quicker and see things develop a lot faster. People want instant gratification—and clients are no different. It is a reflection of wanting to take care of the right here and now deadline crunch, rather than investing and taking care of the long term effects of such design decisions. This can create an atmosphere of trade-offs—price versus quality versus convenience. ”

Keith Bright, designer
Bright & Associates
Venice, California

“ We all have an innate sense of what we like and don't like. Often, though, we can't really say what it is we like—what works versus what doesn't work, or what it takes to make what we don't like become what we *do* like. Without visual skills, we don't know how to fix it. So we start over and scrap what may be good ideas. No time savings. If you need something typed, ask a typist. If you need something written, ask a writer. If you need something designed, ask a designer. If you need a newsletter, ask all three. ”

Betsy Shepherd
freelance writer
Chicago, Illinois

" Keeping desktop computers in perspective as a design tool is a challenge for our industry. Computers open infinite possibilities for the designer, but allow bad design to happen more frequently. If our function is to communicate, we must base that communication on sound concepts and apply good design principles to make that communication effective. I see a lot of work that is aesthetically very nice but says very little. "

Randy Messer
Art Directors Association of Iowa
Des Moines, Iowa

" Desktop publishing is often a solitary endeavor, and it's marketed that way, as if that were a good thing. But what we've lost is the team of professionals who backstopped each other, so the specialists in each field did not make idiots of themselves in areas where they lacked expertise. There were often fights, but they led to creative solutions. The intellectual conflict resulted in better, fresher pieces because two or more heads are better than one. "

Jan White, designer, author, consultant
Westport, Connecticut

" We are challenged to somehow seamlessly continue the excellent creative work we do, learn a new set of skills to integrate the computer into new areas of design media, and master the expertise in production tasks once reserved only for prepress professionals.

John Stewart
Prē magazine

"The computer. What a great tool. Most of us have one on our desks now and are thrilled at how quickly it can do things for us—output copy, manipulate type, alter a page layout. We are also surprised at how quickly time flies while we do 'just one more quick change.' The biggest challenge facing designers now is the *optimum* use, not the *maximum* use, of computers. It is just too tempting to design on-screen what we traditionally did by hand, or delegate to an illustrator, calligrapher, or typesetter. Faster production, numerous drafts, and copious changes are not design solutions. We must be careful not to dedicate too much time to the mechanics of a project and short-change the thought process."

Gail James, Advertising Production Association of Puget Sound, Seattle, Washington

"Computers do not give more leisure time to authors. By removing the sheer drudgery of writing, they encourage them to work twice as hard and produce five times as much."

Arthur C. Clarke
author and scientist
Sri Lanka, Ceylon

"Who was suddenly doing all this desktop publishing? Laypeople. The average computer owner no longer had to contract publishing work to a printer, inspect layouts, edit bluelines, and then wait weeks for the printed results. Now, for many applications, the same steps took minutes. Certainly—and this is an important point about the production of virtual products by customers—desktop publishing demanded a greater participation and understanding by the user. But, as billions of dollars in sales proved, the typical user was willing to make that sacrifice in exchange for control, speed, and lower cost."

William H. Davidow and
Michael S. Malone
Mohr, Davidow Ventures
Menlo Park, California

TELECOMMUNICATIONS AND NETWORKING

13

To send text or image data from one location to another, there are several choices:

- The digital media—be it a disk, CD-ROM, etc.— may be physically transported by courier.
- Data may be sent via telephone lines with a modem. Such non-networked transmissions should be accompanied by a fax for verification. As more and more documents are digital, such transmission is becoming common, saving time and money. Speed and accuracy make it more efficient.
- Digital information may be sent over network lines. This depends on who is networked with whom and the formats of each hardware and software configuration. Networking is an area where the technology is progressing rapidly and changing how professionals work together. Advances that will move the industry further into the information age include:
  - LAN (Local Area Networking)—currently workstations must be hard-wire linked and require a dedicated computer as a server. Collaborative computing, or Groupware, is becoming more prevalent. (See Chapter 14, page 74) Documents can be worked on by more than one user at a time (along with edit control features), annotated with voice or color, and with version control management.
  - WAN (Wide Area Networking)—again, this is still a hard-wired system. Buyers can be linked with suppliers and suppliers can provide maintenance assistance.
  - Wireless networking technology—this will enable data services to be user-dependent rather than location-dependent. The user will not be tied to any location, as capability will go with the user from location to location. For example, with an electronic access badge, a document can follow a worker from office to office.
  - Telecommuting—notebook computers are accelerating advances in both portability and input technology (pen-based or voice-activated versus using a keyboard). This technology will enable more workers to utilize

Edit control in collaborative computing enables the document originator to control how other individuals interact with their document during a real time computer conference. One user may only be able to edit at one time, or all can edit simultaneously as a brainstorming session.

Care must be taken when transferring files either through modem or by disk. Different software and hardware configurations may have incompatible formatting. Technology changes quickly in this area, but it may take several attempts to get files to transfer. The Client should be warned that this is not always a smooth process initially.

In large information system technology, electronic data interchange is accelerating the acceptance of open systems. In the publishing world, hardware and software systems are still proprietary, but that will change in the near future. Open systems in publishing will allow a document to be opened in any software to utilize that particular software's features. Different software packages will also be more linked than they are now, allowing updates in one package to automatically update in the linked package.

computers. There will be an increase in off-site workers, either in the field or at home. This will require a new philosophy of management which is task-specific more than location-specific.

- Video conferencing—video "telephones" where conversations can be live (with many locations at once) or conversations can be handled as video messages. These are more humanized than voice mail, as the recipient can see the person: their gestures and expressions.
- Storage or library links—networks can link the worker to offsite storage facilities or library banks that can offer information and images instantly.

Whoever is transmitting the data is responsible for accuracy and cost of transmission.

MAJOR DECISIONS

- plan project
- solicit proposals
- choose Creative Group
- finalize the proposal
- select the design theme
- approve the design
- review the writing outline
- edit the first draft manuscript
- review the revised manuscript
- review the final manuscript

ORIGINATING CLIENT ROLE

When digital information is sent to the Creative Group, it should:

- contain instructions and a time frame for work to be performed (if not already determined through the planning between Client and Creative Group).
- be accompanied or followed by hard copy, usually faxed or sent by courier.

CREATIVE ROLE

When the Creative Group receives information from the Client, it must:

- acknowledge that the information was received, put into process, and documented.
- notify the Client if the information changes the project parameters or results in extra costs. This should be done before requested work is performed.

PRINT ROLE

Unless otherwise agreed, the Buyer (whoever is purchasing the Print services) pays for transmission charges.

- The Print Group is not responsible for:
  - any errors, omissions, or extra costs resulting from faults in transmission
  - incompatibility between the sending and receiving computers or software.
- The firm sending the data is responsible for the correctness of the received data.

WORKGROUP 14

Today's cutting edge of publishing centers on the ability of individuals to do more digitally and to do it interactively with other workers. It is a blend of technology and people working together. Although the roles of who does what are blurring, the process itself is definable. This process can be placed in a software format that brings together many workers over a network. True workgroup publishing is more than transferring documents from one computer to another, it is:

- a way of people using computers on a network to collaborate towards a common goal.
- several people working on the same document simultaneously by using groupware.
- a system for version control.
- customized for lock-out or role definitions for editing and security.
- track progress and check process completion.
- able to coordinate project and workflow management.
- a means of information sharing through databases and messaging functions.

The essence of groupware is dealing with worker relationships: owner to user, requester to fulfiller, coordinator to members, assigner to doer, proposer to approver, etc. Groupware is abstract and those using it don't need to worry about the system underpinnings. Multiple workgroups can also interlink and overlap, coexisting in a web-work of productivity.

ORIGINATING CLIENT GROUP

What the Client wishes to accomplish through a workgroup will determine the equipment used. Cultural changes are necessary because workgroup tools can't be assimilated into an organization as simply a new way to do what is already being done. These tools will change the way the work is being done. Such changes include:

- role redefinition and adjusting.
- work flow transformations.
- collaboration between workers in new ways.
- developing cross-skill sets to transfer easily from one function to another.

MAJOR DECISIONS

- plan project
- solicit proposals
- choose Creative Group
- finalize the proposal
- select the design theme
- approve the design
- review the writing outline
- edit the first draft manuscript
- review the revised manuscript
- review the final manuscript
- approve visual components

- clearer ways to trouble-shoot and constantly improve and evolve the system.
- setting electronic guidelines and page standards that fit the company's unique way of working.
- setting appropriate time frames and work expectations.

CREATIVE GROUP

The software promotes a greater collaboration between the writers, editors, designers, and producers of documents. Small organizations may work in tandem with other small groups or directly with their clients. The originator of the file, so important in copyright definition (see Chapter 29, page 162) becomes harder to determine. Most documents will originate with the Client Group, each participant adding their component to the project. The Creative Group needs to:

- be clear about what they are contributing and create images and text in their own files for exportation.
- communicate limits of participation upfront.
- help network-mates understand their role in the process, along with the software controls.
- learn various forms of version and annotation techniques to clearly communicate through the evolution of a document.

PRINT GROUP

The production processes will continue to compress the distance between monitor and digital presses. Eventually, the transmittal of a created file will tie in directly to a high-speed, high-quantity output device. The printer will be part of the workgroup and will have input for printing specifications before the project is technically finished. Other workgroup production capabilities will include:

- electronic price estimation through various choice scenarios of press sizes, etc.
- on-line paper specification and availability.
- customization of print orders.
- directly interfacing with distribution mechanisms such as direct mail service, broadcast faxes, and multimedia.

VIEWPOINTS:

COLLABORATION

"Managers tend to concentrate on the expenses of hardware, software, and training. They don't realize they need a fourth evaluation: new responsibilities and new workflow.

We find that technology has completely changed the services of our marketing services group. Many in our bank who used to depend on our services for their presentation needs now do it themselves. Because they handle the execution of what they do, the quality has gone down, but they don't realize this because they fall in love with their own work. Instead of selling or doing other things they are supposed to be doing as part of their jobs, they are playing artist. They are enthusiastic, but not skilled as editors or designers.

So now my area is called on to fix up and rescue many of these presentations. We also don't do as many brochures. Instead, we design formats and templates because technology allows for customization and small-run imprinting."

Christina Nemanic
Harris Bank
Chicago, Illinois

"The changing world of information technology is similar to what has happened over the years to clinical medicine. People once went meekly to their family doctors, who diagnosed their ills, prescribed a therapy and provided it—no questions asked. Today, a physician's diagnosis likely will be greeted by a patient holding two magazine articles on new therapies, who leaves the doctor's office to get a second opinion.

Likewise, technology professionals, once seen as almost mystical information keepers whose centralized department was known as the 'glass house' in popular jargon, have lost their right to ladle out technology as they see fit—no questions asked.

It is a profound and widespread shift in the modern business landscape, wrought in large part by technological expansion that appears to be accelerating out of any one group's control."

Jon Van, reporter
Chicago Tribune
Chicago, Illinois

Gary Carder, President
CarPec Graphics, Inc.
Akron, Ohio

"The weakest link in your work group is the most inexperienced user."

Anne-Marie Concepción
Seneca Design & Consulting
Chicago, Illinois

"When troubleshooting, your credo needs to be: 'God grant me the strength to fix the problems I can, call someone in for the ones I can't, and the analytical skills to figure out the difference.'"

Small Business Report
April 1983

"The DP manager's creativity is often the reason for unwarranted growth. They enjoy devising and then testing new applications. Unfortunately, they rarely propose cost-effective methods of using these applications. A creative, unmonitored computer technician can incorporate new applications and programs to the point where costs exceed value. Stress cost effectiveness to the DP manager by asking: What is the annual cost of running the most-used computer applications for the current year? How are new applications evaluated? Do the actual benefits match the projected benefits? Are limits for governing the use of the system set? If so, how? Are they realistic?

Management must ask these questions and demand satisfactory answers before any new procedures or purchases are approved. The potential saving justifies this attention. Don't be stampeded—remember, most growing companies feel their systems are inadequate, but the root problem can usually be traced to ineffective systems planning."

George Goldsmith, President
The Human Interface Group, Inc.
Wethersfield, Connecticut

"Workgroup technologies are only successful if you understand how they will improve your business process."

“I believe that productivity level has greatly increased in our company because of desk top publishing, proper training methods, managing job descriptions, setting standards, making sure that everybody knows the proper end product and all the steps that are necessary to get there, and to maximize that they are part of that loop. They must know the job that is ahead of them, behind them, and how it all works together. And from an owner's standpoint I am seeing productivity increases that are reflected in the bottom line. This allows me to continue to invest in new technology, to keep people paid well, to send them to seminars, to educate them and bring them along in this new world we've created. We're hit, though, with so much new stuff, what does the production staff do first: learn or produce? Cross training is so important: knowing the jobs of two or three different areas—although you know you don't want to be a typesetter, you must learn to be a typesetter. Although you don't want to be a designer, you must learn what design means to the total project. You don't want to be a scanner operator, you must learn what the scanner operator goes through to get the flow properly into the network.”

Gary Carder, President
CarPec Graphics, Inc.
Akron, Ohio

“A workgroup environment has improved our productivity. We are able to communicate because of the definition of each job. I'm a stickler on procedures. I laid the workflow out precisely and defined each person's position and job. An editor would not touch a picture and change it when they are making edits to their text. The flood of information that has been hitting us has improved productivity if you improve your people and communication skills.

It makes people take on more personal responsibility for their jobs. In the past, you always had someone else to blame. You would send your text out to be typeset and the typesetter made the mistakes. If you had someone shooting a stat for you, and they did it wrong in the darkroom. You had them redo it. But now, if you know what your role is and it is getting defined, then you have to take responsibility for it. I see it as an advantage.”

Cathy Reedy, Art Director
Johnson Publishing Company
Chicago, Illinois

15 NEW MEDIA AND MULTIMEDIA

Some multimedia associations:

Optical Publishing Association
P.O. Box 21268
Columbus, Ohio 43221
614/442-8805

ASCAP
(the American Society of Composers, Authors, and Publishers)
One Lincoln Plaza
New York, New York 10023
212/595-3050

Picture Network International
2000 15th Street North
Arlington, Virginia 22201
703/558-7860
707/312-6210 for membership information

IICS
International Interactive Communications Society
Executive Office
14657 S.W. Peal Bloulevard, Suite 119
Beaverton, Oregon 97007
503/579-4427

Exciting new avenues of communication are opening to reveal a landscape of possibilities. From direct to consumers sitting in their living rooms, to kiosks in retail centers, to screens inside cars, to portable computers, multimedia is infiltrating simultaneously many levels of society. The purpose of a multimedia publication is to present, promote, or explain products, services, or messages. The desired effect on the audience could be inspirational, impactful, or educational.

Definitions

There is a distinction between multimedia and interactive media:

- Multimedia combines text, graphics, video, and sound in presentations and communicates through motion and animation.
- Interactive media is a subset of multimedia, where the viewer can control experience through choices. As a one-on-one medium, uses will be widespread. A great advantage is that recipients can view an interactive presentation whenever they want, and run it at their own speed at their chosen level of detail. The structure is branched, open to selection and different for each viewer.
- Blended media combine several communicative vehicles into a campaign where a conceptual design theme is applied to print, interactive media, broadcast, and display. Clients will gain from such consolidation both in market consistency as well as economically when elements are shared between different uses.

Applications

The new media are meant to create an experience through a process that is enriching. In electronic publications, the medium is the screen, combining text, images, movement, and sound. There are many new ways to communicate:

- information kiosks such as in stores or corporate lobbies.
- electronic magazines customized by subscription.
- interactive advertising via disk or CD-ROM.

- on-line publications through a network.
- electronic books.
- manuals and directories distributed on CD-ROM that would be expensive to update in print.
- instructional devices that give "readers" choices.
- augmented presentations such as for sales or education.
- a new form of "brochure" that is controlled in depth of content by the viewer.
- dictionaries and other reference materials.
- publications for portable computers.
- catalogs and interactive ordering or shopping.
- newsletter and bulletin faxes.
- broadcast faxes.

New media have a range of output methods:

- for small files, floppy disks.
- for large files, distribution can be through
  - Syquest disks.
  - CD-ROM.
  - Videotape.

The growing flurry of development and activity is untempered by incompatibilities between formats and lack of standards. Manufacturers are in a race to produce the best and the most comprehensive new possibilities to reach an audience.

From development to users

The market for these new forms of communication is growing, and so is tremendous competition between the providers of development services. In many ways, development is running ahead of demand; a medium waiting for its market to catch up. Currently there are many advantages and disadvantages to the new media:

- Advantages:
  - can expand the use of existing images through the repurposing of (finding new uses for) content.

For designers transferring from print design to multimedia design, the new skill levels needed can be overwhelming. Even for professionals staying within their specialty, so many new developments require constant learning. The busy worker must have ways of coping with change. Each individual must recognize that they will never be caught up on:
> magazine reading.
> advances in hardware and software.
> items on a "to do list."
> correspondence.
> phone calls.
> conferences to attend.
> software idiosyncrasies.
The best thing to do is plod along with trying to be ahead of everything, have lots of friends, and do what most needs to be done.

- time independence.
- involves more than one sense (sight and hearing) of perception.
- involves the user through giving control and navigation flexibility.
- can be inexpensive to deliver.
- may integrate other media, such as still pictures, animation, and sound.
- can make complicated information easier to grasp.
- can be unusual, making promotions stand out.

- Disadvantages:
  - electronic portion dependent on viewing device, often not portable.
  - only one user at a time can experience.
  - expensive in design and development.
  - gaining permissions can be very complicated when negotiating with several industries for audio, graphics, video, and text.
  - hardware and software may be incompatible.
  - lack of standards.

MAJOR DECISIONS

plan project
solicit proposals
choose Creative Group
finalize the proposal
select the design theme
approve the design
review the writing outline
edit the first draft manuscript
review the revised manuscript
review the final manuscript
approve visual components

ORIGINATING CLIENT

The astute client is not seduced by media technique, but is focused on the purpose of the organization's publications. To the educated manager, new media are valued if they present new ways to reach goals. The medium does not change the purpose of the message but rather the form of the message. To utilize multimedia effectively, the Client must balance:

- business goals and expectations.
- audience characteristics and interests.
- background information and current image resources.
- potential information sources, careful of copyright (see Chapter 29, page 162).
- available content with appropriate new content.

CREATIVE GROUP

There are more professionals involved in the process than there is in print publishing. The team must also include these additional individuals:

- Producer
- Director
- Production staff
- Audio and video professionals

New media offer possibilities for new forms of visual language. The hardware and software are growing rapidly with capacity and complexity. The creator of multimedia projects should:

- **Plan**
    - be very clear about Client goals and expectations before beginning a project.
    - become content experts, understanding resources, libraries, ownerships, and legalities.
    - have a team approach, for no one person can be an expert in all areas of multimedia.
    - organize information and use as much original material as possible, securing all necessary permissions for existing material.
    - utilize the simplest software early in the project while assembling components and then progress to the more complicated software later in the project.
- **Design**
    - begin with flow chart, outline, and storyboards.
    - use tree-formatting for interactive design to make sure that the viewer does not miss any essential points in the possible choices.
    - determine hardware, software, and training needs for project.

- design interface development and method of viewer interaction.
- scan or convert all graphics and visual materials to needed formats.
- clean up graphics for multimedia viewing, which has different visual requirements than print or presentation graphics.
- roughly score and determine flow of project.
- work on various segments and design visual screens, pace, choices, and integration.
- add audio voice track, working with voice talent.
- synchronize animation to the audio track

- Development and production.
  - combine all images, text, and sound
  - develop sequence along branching (for interactive).
  - design menus and transition mechanisms.
  - arrange for testing sources and feedback of several initial users.
  - debug and perfect the flow and sequence of project.
  - integrate with print materials, packaging, and delivery methods.
  - work with those who handle distribution for feedback methods.

PRINT GROUP

This is not really a print group anymore, but would more effectively be called an "Output Group." Many Imaging Centers are expanding their services from handling high-resolution output for print to computer-to-video production. Those companies that have arisen from the video side of the industry also can provide editing services, charging by the hour. The division between the groups dissolves under collaboration.

VIEWPOINTS:

BLENDED MEDIA

"Much of design specialization is changing because new tools allow for a broader range of media applications. The desktop computer may never take the place of a full video-editing console, but the basic tools for rough editing and creative assembly are already available, and are as easy to use as drawing packages.

But with new media, it will be necessary to learn complementary skills: those that mimic our traditional tools, and those needed for multimedia—for instance, sound and motion editing. Because the complexity of work will likely bring to the designer's domain overlapping media, knowing more about these interweaving disciplines will be necessary. We cannot be experts in all technologies and applications, but we can begin to understand how to creatively apply the basic language of each to our work."

Alyce Kaprow
computer graphics consultant
Newton, Massachusetts

"Graphic designers have to look beyond the printed page and start investigating multimedia. Most clients will soon (sooner than we think) have a CD-ROM player and will want their corporate vision displayed on computer screens with animated imagery. They will need to reach this 'techie' audience with more than ink on paper."

Stan Evenson, designer,
Evenson Design Group
Culver City, California

"Multimedia authoring requires the language skills of a writer, the creativity of an artist, the management abilities and sensitivity of a film director, the determination of a project manager, and the patience of Job."

Joel Orr
Orr Associates, Inc.
Virginia Beach, Virginia

Bob Croach, designer
JHT Multimedia
Winter Park, Florida

“When you're starting out, whatever capital you think multimedia is going to take, multiply by 2.5.”

Steve Rosenthal
writer and video producer
Berkeley, California

“Many document-management and e-mail systems concentrate not just on the form of documents but on the way they flow through a workgroup. With electronic distribution, 'you also change the way the information flows at the workplace,' says BIS's Tom Ashley. In most cases, Ashley reports, electronic distribution is faster. Documents no longer have to wait for printing or for transfer by internal and external mail, and virtually unlimited copies can be sent out simultaneously. Electronic publishing is also more flexible—more people tend to get the information, and they get it with fewer routing stops, even over long distances. 'The main result is higher efficiency,' he says.”

Bill Lorenz, designer
Entertainment Technology
Orlando, Florida

“The value of multimedia is that we can communicate abstract concepts—different objects, shapes—and people can watch. We can let the media do the presentation while you spend your time thinking about how well your client is perceiving what you want to be presented.”

Syllabus, April/May 1992

“Multimedia can best be defined as various combinations of text, graphics, sound, video, and animation that are controlled, coordinated, and delivered on the computer screen. Multimedia also implies interactivity where the user is actively engaged in the presentation of information, and is not just a passive observer of a fixed procession of sights and sounds.”

“ Multimedia isn't. Multimedia is about multi*sensory* communication through a single medium: the digital file. A slide show accompanied by audio tape and film—*that's* 'multimedia.' Computer files with text, sound, animation, video, and hypertext links—they are multisensory *mono*medium expressions.

It's not just the 'multi-' part with which I take issue. A medium is by definition a means—not an end. But McLuhan's dictum still holds: The medium IS the message. Multimedia today is like a talking dog; it is not what the dog says, but the fact of its talking, that commands our attention.

In truth, multimedia is a place-holder, a temporary substitute for what we really want: direct contact, immediacy. No medium at all. Intimacy with the subject—our designs, our writing, and those of others. ”

Joel Orr
Orr Associates, Inc.
Virginia Beach, Virginia

“ One of the things going on at WGBH is something we call multiversioning. It is the idea of producing a lot of different products in different media from one basic idea. Rather than saying 'OK, we have the television program, now let's make the book, now let's make the interactive video disk, now let's make the CD-ROM,' it's more with this idea: 'Let's think about what simultaneous media we want to create.' When you start thinking of a project that way, there's a lot to consider in terms of design and production from the start. There is the potential for the designer to have a much larger role both in terms of conceptual work and what's going to be the proper medium for his idea. ”

Wendy Richmond, co-director
WGBH-TV Design Lab
Boston, Massachusetts

> “Effective training can't occur until the trainer understands the learning process. Three core factors in learning are the senses, repetition, and exploration. Generally, the more senses you can involve in training, the more learners will retain. Demonstrations, explanations, discussions, supplemental reading, and hands-on experience all combine to produce retained knowledge in the student, and it's important to include these activities in every computer training class.”

Tori Coward, president
Tangent Computer Resources
Dallas, Texas

> “If electronic documents are ever to become more than disembodied snapshots of paper pages, we need new conventions for how such documents are structured, represented, parsed, indexed, and distributed—and possibly a new aesthetic as well. But while just about everyone agrees on that principle, more than a dozen overlapping, conflicting, and complementary approaches are competing for acceptance.”

Steve Rosenthal
writer and video producer
Berkeley, California

> “'With multimedia, we are still not there yet; we still don't have that critical mass.' said Microsoft Corp. Chairman Bill Gates. However, Gates predicted that the technology will be mature by the end of the decade, propelled forward by programs such as book players and computerized magazines that are inexpensive and easy to use. He also said future versions of the Windows graphical environment will have multimedia built into them. Multimedia extensions to Windows are available today.”

James Daly, News Editor
Electronic Entertainment
San Mateo, California

PRODUCTION

OUTSIDE SERVICES AND OUT-OF-POCKET EXPENSES

16

Outside purchases by the Creative Group or Print Group to fulfill a Client's order may include services such as photography, illustration, or printing. Out-of-pocket expenses include materials, messenger services, deliveries, and transportation.

Any party who purchases "third-party" services assumes responsibility for that service. "Responsibility" encompasses the following:

- accuracy
- completeness
- timeliness
- financial issues
- legal issues.

The Buyer pays for all acceptable work—work that meets all defined Buyer parameters. The Buyer doesn't pay for unacceptable work. A Buyer cannot claim work as unacceptable if it is actually put into use.

Every Group is responsible for the services and materials within their purview. For example, the Writer and Editor as originators of the text are responsible for proofreading; the Printer is responsible for paper, and each member of the Print Group is responsible for the quality and longevity of the media they produce.

ORIGINATING CLIENT ROLE

When engaging the Creative Group, the Client should understand

- which services the Creative Group will provide and which, if any, the Creative Group will purchase from a third party.
- service capabilities vary from organization to organization.
- that it may be an advantage to have the third party bill the Client directly. The Client will save mark-up fees added by the Creative Group for handling the financial responsibility (see next page).
- the Creative Group shares in purchasing responsibilities with the Client. They share in the satisfaction of work performed, especially in the area of printing, unless the Creative Group agrees to take on this responsibility fully.
- the Creative Group approves all invoices according to agreed fees, even when the Client is billed directly.

MAJOR DECISIONS

- plan project
- solicit proposals
- choose Creative Group
- finalize the proposal
- select the design theme
- approve the design
- review the writing outline
- edit the first draft manuscript
- review the revised manuscript
- review the final manuscript
- approve visual components

Each Creative Group has their own mark-up policies when handling the financing of outside purchases. Industry standard is generally 15%.

CREATIVE ROLE

For each function performed or purchased directly, such as writing, design, text entry, proofreading, prepress, buying paper, and printing, the Creative Group takes on full responsibility—including financial—to the Client in the event of any problems, unless other agreements are made.

It is the Creative Group's responsibility to:

- describe to the Client its mark-up policies. This includes shared responsibilities when invoices go directly to the Client.
- describe any policies for direct billing of outside services, especially when such fees are greater than the design fees.
- put together the most appropriate Print Group, holding them responsible for their roles, including financial, according to agreements.
- coordinate, manage, and communicate with the Print Group. The Client pays the Creative Group for this service:
  - through mark-up fees that have been agreed upon in advance.
  - through supervision fees if the Print Group bills the Client directly.

PRINT ROLE

Unless otherwise agreed to in writing, the Print Group will

- mark-up any of their outside costs, such as paper, inks, die-cutting, foil stamping, film, and other materials, which are usually part of their quote.
- define these mark-up fees in advance.

VIEWPOINTS:

INVESTMENTS

"Distribution is one of the most important but overlooked things you'll run into. If you're looking to develop a custom piece of multimedia for a single customer, you could care less about distribution. But if you're going to go out and use an existing CD on dinosaurs, add hypertext, and turn it into an educational program, and get that distributed to schools, now distribution becomes a problem. You've got to know *how* to do it, what the sources are, get to them, and it's not cheap. You have to pay up front and see the revenue downstream."

Bob Croach
JHT Multimedia
Winter Park, Florida

"Multimedia is a brave new world for the entrepreneur. Production costs are substantial — averaging in the range of $300,000 to $1,000,000 per title. While the percentage of the population with CD-ROM players is increasing rapidly, effective distribution is a formidable challenge. The capitalization required makes multimedia less like publishing and, perhaps, more like film. No doubt some producers will thrive, but the risks are great."

Tad Crawford
Publisher and Writer
Allworth Press
New York, NY

17 PAGE COMPOSITION

Page composition is the creation of documents that contain all text and graphic elements assembled into page layouts. Using traditional methods, pages are keylined, which is manual composition resulting in a pasteup. With desktop publishing, combining elements is electronic and a number of intermediate proofs may be created. These proofs provide the Client and the Creative Group greater flexibility in the development of final documents without much of the time and cost associated with traditional methods.

The proofs that precede final documents include:

- preliminary client proof: writing and visual images formatted by the Page Composer into pages, showing how the elements fit. This is rough, for many items are unfinished and only for position. The Client should read this proof thoroughly for any needed adjustments.
- corrected client proof: made by the Page Composer from the Client's review and then refined into the final document.
- final low-resolution proof: prepared by the Page Composer as the last step before outputting the file in prepress for printing. The Client needs to check all the elements, proofreading, etc., and initial and date the hard copy to signify approval of work completed (or with indicated changes). Beyond this point changes may entail extra time and fees.

MAJOR DECISIONS

plan project
solicit proposals
choose Creative Group
finalize the proposal
select the design theme
approve the design
review the writing outline
edit the first draft manuscript
review the revised manuscript
review the final manuscript
approve visual components
approve preliminary Client proof
approve corrected Client proof
approve final low-resolution proof

ORIGINATING CLIENT ROLE

The best time for the Client to specify any needed adjustments in the writing, statistics, content, or design, is before page composition begins. The Client and the Project Manager supervise these phases with the Creative Group.

Before page composition begins:

- the final manuscript must be approved by the Client. All decision-makers should initial their comments on the revision manuscript. The Client incorporates these comments onto the hard copy of a single revision manuscript and provides it to the Creative Group as the final.
- the Creative Group then prepares intermediate proofs for the Client's approval.

As page composition is completed:

- the Client receives final low-resolution proofs to review and sign before the project proceeds.
- the Client may receive electronic files from the Creative Group, rather than page proofs. This occasionally happens when
  - the Client has in-house production and the Designer provides publication templates (electronic page grids and style sheets for typography).
  - the Client becomes the Buyer for final output (see Chapter 18, page 104) and printing. This is most effective when the Creative Group is available for consultation.

Many publishing projects, due to the birth of new media possibilities, are not intended to be printed. The disk becomes the carrier of information. Three areas for new publishing are

> desktop presentations—interactive documents, where information is scripted into various scenarios. Both HyperCard and MacroMedia Director allow the user to choose information from menus. To be sure that the audience receives an overview of the topic, a printed brochure may accompany the disk (integrated publishing).

> fax publications—for a limited, yet speedy, circulation.

> electronic speaker support—for slides, overheads, and video.

CREATIVE ROLE

A number of final check-points and presentations of proofs to the Client are needed in the development of documents. With the Art Director and Project Manager as the core project supervisors, communication should be a smooth process.

- The Art Director or Project Manager:
  - makes sure that the Client has approved the final design.
  - receives final approved manuscript from the Client.
  - supervises the preparation of visual resources—photography or illustration.
  - supervises the Page Composer to make sure page composition adheres to the design.
  - is responsible for quality control of elements, including completeness, budget, and schedule.
  - provides estimates to the Client concerning any changes made that alter parameters. (Any changes that do not alter project parameters are provided at no fee.) The Art Director must be sure to discuss any extra fees for changes. If the Client is unavailable for ap-

A project schedule may start out compressed or become compressed for a number of reasons. As the Page Composer finishes, there may be last-minute changes, but there may not be time in the schedule to stop for estimating those changes. The Creative Group is faced with jeopardizing the Client's deadline or with financial risk by making changes without the Client's approval for additional expenses. To avoid this common trap, the Creative Group must clearly communicate with the Client that extra fees may be added for any changes made during final production. (Of course, it is necessary to clearly communicate with the Client what steps are considered part of production) A Client has approval over how they spend their budget.

proval, then the Client accepts the best judgment of the Creative Group, unless the Client agrees in advance to hold up production until approval. If the Creative Group does not discuss extra fees with the Client when the Client is available, the Creative Group undertakes changes at their own economic risk, for the Client may want the option of which changes are worth extra fees.

- checks over the final low-resolution proof.

- The Page Composer
  - works directly with the Art Director and Designer to produce pages according to the approved design.
  - receives instructions and materials, which include the approved designs and comprehensives, electronic manuscript, and the visual elements and images.
  - prepares the electronic project files: formats all text, places text into document file, combines all text and graphic elements, and builds pages.
  - is conversant in the highest quality, lowest cost production methods and can recommend which steps to complete in-house and which to assign to the Print Group.
  - interfaces with the Print Group and is aware of all agreements, specifications, and technical procedures.
  - meets with the Print Group to determine the best way to prepare documents prior to providing finished files.
  - places date and file name on all output.
  - prepares preliminary proofs first for the Art Director and Designer, then for the Client. After the Client review, the Page Composer prepares a corrected Client proof and final low-resolution proofs for the Art Director, Designer, and Client.

When page composition software was first introduced, each package had a fixed feature set. But as capabilities have expanded and systems have become more powerful, the software is becoming modular. Users can customize their configurations for their own needs, by way of Aldus Additions, Quark Xtensions, etc. Future generations of software will become even more seamlessly compatible: Documents will be created in one version, then modified in another application to use its own feature set.

- prepares files for prepress, such as adding crop marks or setting up overlaying methods.
- handles color separations, where appropriate. If the Page Composer creates the separations, the Page Composer is responsible for how those separations will function for the Printer. For example, the Page Composer becomes responsible for register marks, trapping, and other built elements.
- provides to the Print Group all final electronic files, complete instructions (such as system configuration, software versions, font requirements, and output specifications), final low-resolution proofs, any fonts the Print Group doesn't have (to be used only on that document, see Chapter 30, page 171), separated proof pages or composite proofs for any multi-colored specifications, and any other required components.
- keeps backup files of all electronic and proof documents in case of later possible Client changes.
- is available by phone to answer any questions the Print Group may have, especially when new processes are being used.
- sets up communication between Production Coordinator and Printer.

Creative Groups have found they can save a lot of time and money by "partnering" with the Print Group long before page composition begins. If the Print Group sees the concepts, they may suggest cost and time saving ways to produce the ideas. This should continue throughout the project.

Many production decisions may influence page composition and output preparation. Every press has different tolerances for color registration and film densities. Different papers require different halftone dot structures and ink consistencies. Also, with color printing, color can vary greatly from the monitor, where a document was created, to the printed piece.

PRINT GROUP

The Production Coordinator or the Prepress Specialist and the Printer should be available to answer questions throughout the process. They serve as resources to the Page Composer, ensuring that the pages are constructed in a way that will go through the imagesetter efficiently (See Prepress Chapter 18, page 102). In this way, they are a resource and work in partnership with the Creative Group.

The most common output problems reported by Imaging Centers are incomplete instructions, pages with unnecessary elements, and font incompatibilities. Such problems can be avoided if the Imaging Center advises the Page Composer on appropriate preparation and receives a test page before page composition is completed.

VIEWPOINTS:

ELECTRONIC TOOLS

“A big part of the ingenuity of design is coming up with ways to get things produced.

The client-designer relationship should allocate ample time and consideration to this topic. It can take as much (or more) time as conceptual phases to sift through the nuts and bolts of how something is going to get to its finished form.

The process can be frustrating. Finding ways to produce things that are inherently different is not an easy task.

Keep in mind that it is every bit as creative and every bit as important.”

Richardson or Richardson,
Hopper Paper Company

“It's like housework. Do a little bit every day, and you keep ahead of things. Let it go for a few days, and it starts piling up. Eventually it gets bad enough that priorities must shift. Something must be done.

The same is true with computer output and files. When you get to the point where output is everywhere, diskettes are everywhere, and you see dozens of names on your hard disk that you recognize but can't associate with faces, it's time for action.

Keeping track of output and data is really no difficult task. It's just a matter of a little organization and a little time each day to adhere to that organization.”

John Ivory, writer
Chicago, Illinois

“Designers should do it all. Printers *love* change orders. (Laughs) Actually, we recommend that our clients do not do traps, etc. but if you are going to do it be aware of the responsibility you assume.”

Kurt Klein
Customer Support Services
Graphic Arts Center
Portland, Oregon

“Your corporate responsibility goes beyond choosing the proper supplier. In no instance, whether selecting a specialist or generalist, should you release control or influence over the process that leads to the proper answer. You cannot simply describe your problems and then remove yourself from the stages leading to solutions. Feedback and direction must be provided so that progress remains on a true course or is repositioned before time and effort are wasted. You must develop the original objectives, review the initial concepts, eliminate and refine, then when you finally judge the outcome, the rationale and ultimate solution should tie into those original objectives.”

Anthony A. Parisi
Director of Corporate Design
General Foods Corporation

“In the rapidly changing world of desktop technology, strong vendor relationships are paramount to success. Vendors must stay current with a barrage of technological advances, or be buried with the past. We users must draw upon their specialized knowledge to build our future and direct our businesses to survive and excel in this constantly evolving business climate. The most successful vendors stay ahead of technology acting as R&D centers to help their clients stay ahead.”

James Hicks
Director of Print Services
Valentine Radford Inc.
Kansas City, Misouri

“Printers often clean up files and don't tell the designers or page composers what is wrong with the file. Not telling them does not help the situation. But printers fear losing their job if they share their knowledge.”

Clint Funk
White & Associates
Northbrook, Illinois

“ Lithography is not trivial. It is insulting that designers think all they need to do is buy software and they can do a printer's job. Remember: a journeyman's skills require years of training; on-the-job training is the best teacher; only experience yields expertise; only expertise yields results. Similarly, it takes more than drawing software to make a good designer. Remember: good design is a highly developed talent; design is visual, not mechanical; design requires talent, not template. Both side must recognize their limitations *and* their opportunities. Both should openly pursue a synergistic relationship with the other. ”

Herb Paynter, President
ImageXpress
Lawrenceville, Georgia

“ In today's business climate many of the functions provided by specialized vendors have been combined into the desktop operator's job function. These continually changing responsibilities have left operators to fill gaps previously provided by the vendor. Before each specialist contributed knowledge and experience to the finished piece. By shortening the process we have deleted many vital steps. These functions and others, insure that our agreements with vendors facilitate accuracy in the final product and are best done through establishing a relationship. We must agree on where the responsibility for errors between disk output and the final product will be eliminated. We must also agree on what proofs are to be seen and in what form. These agreements must be made at the beginning of projects, not during or after. Agreements guarantee consistency and accountability that can be sustained into the future. ”

James Hicks
Director of Print Services
Valentine Radford Inc.
Kansas City, Misouri

PREPRESS 18

The Page Composer sends the electronic files and the low-resolution proofs to the Print Group for prepress. Advance planning directly affects quality and efficiency. Adequate time for prepress activities is critical. When under unrealistic deadlines, the price goes up (rush charges) and the quality goes down (not enough time to check elements and finesse output preparation).

Prepress includes

- scanning for color and black and white images. Scanning is the digital capture of photographs, illustrations, or line art. These files may then be incorporated into electronic pages or output separately. Some scanning is done at the page composition stage—most notably low-resolution black and white scans of photographs to serve as position images to be replaced by high-resolution color scans when in prepress.
- creating "print files," which are a special way to capture all the information needed to output directly to the imagesetter, including graphics and fonts. Whoever sets up the "print files" is responsible for software print settings, system configuration (such as font installations), and any consequential improper output (see page 103).
- preparing color separations for backgrounds, screens, and other images.
- preparing color separations for multi-color projects. Color photographs are converted into the four process colors (each requiring a separate piece of film). Final output is color-separated film.
- imagesetting: the file is sent to a high-resolution output device, which generates final paper or film output. (This involves several steps in itself).

The Imaging Center provides final output according to Buyer's specifications.

- **High-resolution paper output** is needed when traditional keylined boards are appropriate, usually when a combination of electronic and traditional elements are used. Paper is also chosen when the project doesn't fit the limitations

AISB (Association of Imaging Service Bureaus) is an international trade association for individuals utilizing imagesetters. Members include Service Bureaus, Trade Shops, Printers, in-house imaging departments, and suppliers. For information, write to AISB, 5601 Roanne Way, Suite 605, Greensboro, NC 27409 or call 800/844-AISB.

There is a common expression in electronic publishing for monitor viewing called "WYSIWYG," which means "What You See Is What You Get." When dealing with "print files," it would be accurate to say that "What You Send Is What You Get."

Typographers International Association (TIA) is a 75 year old organization beginning with typographical professionals and not expanding into electronic publishing. Their emphasis is on the craft of the beautiful page (printed or electronic). For information, write to TIA at 84 Park Avenue, Flemington, New Jersey 08822 or call 908/782-4635.

As many corporations expand their in-house capabilities to include prepress methods (such as high-end workstations, color printers, or imagesetters), their need for outside Imaging Centers still remains. The technology evolves so quickly, it is not cost-effective for the corporation to invest in the latest technology. The future role of the Imaging Center may be to function as a research and development site, support center, and consulting center for the Buyer.

of output technology (such as an oversized piece), or when last-minute changes are expected (which can be handled as strip-ins rather than requiring new film). The trend is moving away from paper output as technology allows more files to output directly to film.

- The Client approves boards before final film is made.
- The boards go to a Trade Shop or a Printer for film preparation.
- The Client owns the boards, and the Printer owns the film (see Chapter 28, pages 157 and 160).

- **Film output** is generally produced when a project is entirely digital, when color separations are needed, or both.
  - Film allows higher image quality than paper output, for paper output must then be photographed to create film for the Printer—which adds another optical generation, reducing sharpness.
  - Technological advances are allowing more and more files to output directly to film.
  - There is no intermediate step for Client approval, as with art boards.
  - The Client owns the film (see Chapter 28, page 157).

The Printer, Color Separator, or the Trade Shop places traditionally keylined boards into a copy camera to make the film needed to expose plates. The copy camera is a large high-quality reproduction camera (similar to a stat camera) where the boards are held under pressure as they are photographed for film conversion. Keylined boards or illustrations done conventionally are also called reflective art (see Glossary).

New technology can allow electronic files to image directly to the plate—without film—sometimes right on the printing press. Through laser technology and advancements in the printing plate, this has become the future direction of printing. Quality will take time to develop. As with any new printing technology, it is first applied to projects where quality is not critical. But eventually, this technology should reduce costs, especially with color printing.

ORIGINATING CLIENT ROLE

Many Clients do not know very much about prepress. Once they sign off on the final lasers of a project, they often feel that they are done with it. However, so much of electronic publishing comes together at the prepress stage which then makes the project possible through ever-growing sophistication in imaging technology. The Client has several key responsibilities during prepress and:

- should be aware of, and approve, the schedule and any fees for this phase.
- allow adequate time for prepress to ensure quality. For new methods, it is best to allow twice as much time for output as estimated.
- should understand an overview of the process and work closely with the Creative Group to develop accurate expectations.
- should also keep track of the film and make arrangements for its receipt or storage (see Chapter 26, page 146).

MAJOR DECISIONS

- plan project
- solicit proposals
- choose Creative Group
- finalize the proposal
- select the design theme
- approve the design
- review the writing outline
- edit the first draft manuscript
- review the revised manuscript
- review the final manuscript
- approve visual components
- approve preliminary Client proof
- approve corrected Client proof
- approve final low-resolution proof

CREATIVE ROLE

The Art Director is responsible to the Client if any problems arise. Therefore the Art Director should check references and samples in advance when utilizing a Print Group for the first time.

Collaboration between the Creative Group and the Print Group, *before* the Page Composer prepares electronic files, allows a faster finishing process. Send a sample page in advance of the project. Many Page Composers make this mistake when they send a project with an urgent deadline to the Imaging Center without a previous test or input from the Page Composer.

The Creative Group should request from the Print Group advice on building their pages:

- achieving visual effects, such as the handling of typography, backgrounds, vignettes or blends, borders, etc.
- combining several image formats and methods for setting up these files. These may include the treatment of photographs, illustrations, and other elements created in software other than the page composition software that will be inserted into the pages.
- photographs require special screen information settings to ensure compatibility with the Image Center's output devices and the Printers' specifications.
- when to combine traditional methods with electronic methods. Although the technology is advancing rapidly, there are cases where traditional techniques are faster and more economical.

When sending electronic files to the Imaging Center, the Creative Group should see that the Prepress Specialist understands the Printer's specifications, such as paper and press tolerances.

- This communication will enable the Prepress Specialist to prepare output appropriately.
- If this communication is not facilitated, the Creative Group may have to absorb extra costs if the Printer cannot utilize the output provided.

Criteria for selecting an Imaging Center:

- Match capabilities with needs. If a project requires typographically related services (such as formatting, styling, or editing) make sure the Imaging Center has experience in this area.
- For high quality projects, select an Imaging Center that has both traditional and electronic experience. In this way, the best techniques will be employed for each project's needs. If the project requires color separation, make sure the Prepress Specialist is a color craftsman.
- Check the imagesetting or color separating equipment used. Review samples of projects that most closely relate to the project needed.
- Check compatibility of equipment between the computer system where documents are created and the system of the Imaging Center.
- Check to make sure the Imaging Center has the appropriate fonts available.

Electronic software becomes more and more sophisticated in handling prepress techniques. Trapping, the printing of color images next to one another (such as colored type on another color background), is especially tricky when done on a desktop system. There must be the proper color overlap to avoid white spaces between colors when the project prints. Different presses have different tolerances for trapping, so if the Page Composer wishes to undertake this function, the settings should be verified by the Print Group.

There are two ways for the Page Composer to provide files for imaging: "live" documents (created within a page-composing or drawing application), or "print files" (such as Encapsulated PostScript files). The Production Coordinator should advise on the best approach for the particular project.

- In providing "live" document files, the Page Composer
  - sends the electronic files complete with appropriate graphics and images to the Prepress Specialist.
  - gives instructions to the Prepress Specialist on preparing "print files." Unlike a "print file," the document file does not necessarily carry fonts within it. The Imaging Center is required to own the appropriate fonts and applications for outputting "live" files. The transmission of fonts to the Imaging Center is generally prohibited, except in "print files" (Chapter 30, page 171).
  - there may be some differences between results from the native system (where the file was created) and the foreign system (where the file outputs).
- In creating "print files," the Page Composer
  - chooses the most appropriate settings, such as for line screens, page orientation, percentage of enlargement or reduction, or crop marks.
  - sets up the pages for the desired kind of output.
  - has more control and may save time and money in output.
  - is responsible for how the files perform. If the files do not image appropriately, the Page Composer absorbs any additional costs associated with correcting errors and re-imaging the project.
  - does not require the Imaging Center to own the needed fonts because the fonts are embedded in the document. These files typically can't be opened or edited by the Imaging Center. Therefore, special care should be placed on all settings necessary for these files to image properly.

AISB (The Association of Imaging Service Bureaus) recommends that Imaging Centers inform their customers who provide "print files" to give instructions for desired output and always select "positive" and "right reading" in the application's print dialog box before creating "print files." If the Imaging Center knows the settings are consistent, then the Prepress Specialists know how to adjust their settings to get the desired results. For example, the combination of negative settings in the "print file" and the imaging equipment set at negative yields positive film. Because the most common error in imaging "print files" is getting the wrong kind of film output, the Page Composer should check with the Imaging Center to verify the correct settings before creating their own "print files."

Many files that Imaging Centers receive have "fatal errors"—the document will not open at all. Other files may have minor errors that cost time and money for extra preparation. The Page Composer can avoid most of these errors and costs by discussing the document composition in advance with the Production Coordinator.

- ▻ has fewer chances of error because these electronic files are created on the same system where they originate. These files will typically be downloaded directly to the imagesetter and will image according to the original system's software and hardware settings.

When receiving final output from the Print Group:

- the Page Composer and the Art Director check over the quality and consistency of the final output and materials.
- The Page Composer receives the original disk and support materials back from the Print Group and returns appropriate materials to the Client.

PRINT ROLE

Generally, the Page Composer is the Buyer of imaging services. In some cases, the creator of the electronic file turns it over to the Client Group, who then becomes the Buyer. In such situations, the Page Composer should be available for consultation.

The Production Coordinator handles output and printing preparation. This includes:

- giving advice to the Page Composer on the best way to build and prepare pages, including file preparation for graphics and text (see pages 102 through 103).
- providing guidelines and forms for the Page Composer or Buyer to fill out, with output specifications needed to ensure film and press compatibilities.

Upon receipt of the electronic file, the Production Coordinator:

- has no liability for being able to open, evaluate, or read the file.
- checks project materials and readable electronic files for completeness of
  - software and hardware compatibilities.
  - imaging requirements and instructions.
  - final low-resolution proofs.
  - inclusion of all appropriate graphics needed to image the file.
  - availability of appropriate fonts.
  - print settings according to the Page Composer's guidelines.

Many Imaging Centers are instituting a "pre-flight" department that handles incoming digital files for prepress. A Prepress Specialist is assigned to checking incoming files immediately for imaging ability. In larger facilities, there is a full-time staff whose sole responsibility is checking files and communicating potential problems or errors.

- notifies the Buyer *immediately* and requests additional instructions if
  - instructions are incomplete.
  - file setup is inadequate.
  - errors are detected.
  - preparation is not according to the Print Group's guidelines.
- accepts the project at their own discretion if there are questions and the Page Composer or Buyer are not available.

Imaging errors can be categorized as follows:

• Industry-wide errors: Imaging problems that occur because of poor PostScript code that prevents files from imaging properly. These are very difficult to explain and are not the Imaging Center's or the Page Composer's fault. Correcting these errors are part of the Imaging Center's overhead of doing business. Similarly, if the Page Composer's computer malfunctions, it is the Creative Group's responsibility to repair damaged or lost files. These evolve and change as capabilities mature.

• Software errors: These errors may be inherent in a software application package. The vendors to the Imaging Center have a disclaimer agreement which prevents them from being held liable. It is the responsibility of the Imaging Center to deal with these errors, and they should report them to the application vendors who can correct them in new software releases.

• Page Composer errors: The document originator builds pages improperly. These errors can include font incompatibilities, buried images, or incorrect settings. It is the responsibility of the Imaging Center to request corrected files from the Page Composer, who must absorb the expense of correction and additional imaging time.

• Imaging Center errors: These errors are caused by improper equipment set up or maintenance, or by not following the Buyer's instructions. They are corrected by the Imaging Center at no cost to the Buyer. All Imaging Centers should utilize density control devices and have quality control procedures in place.

The Imaging Center can handle a received project in one of two ways:

- run the project "live," which means opening the files in the original software and then creating the "print files."
  - The disadvantage is that opening the files on a different computer, other than the one where the files were created, can introduce errors because of the different system configurations. (Native versus foreign systems, see page 103.)
  - The advantage is that the Imaging Center can select the proper settings and has more control for the way the files perform for output and printing.
  - The Imaging Center is responsible for the film's printing performance.
- guide the Page Composer in setting up the "print files" on their own system and then download the files to the imagesetter.
  - The disadvantage is that the Imaging Center has less control of how the files image.
  - The advantage is that when the "print files" are prepared properly, there are fewer chances for errors.
  - The Page Composer is responsible for the file performance when preparing "print files" without the advice of the Imaging Center.

When working on the project, the Imaging Center:

- should not perform additional work to image an electronic file without approval from the Buyer. This additional work may include
  - restructuring the file to ensure printability.
  - translating files or graphics (i.e. TIFF and EPS files—see Glossary) from one form to another.
  - correcting print errors.
- utilizes the electronic file as the road map for combining text, images, and other graphic elements to construct the document for imaging. Electronic files can be
  - transferred directly to the imagesetter.
  - transferred to a higher-end workstation for photographic, image, or color manipulation.
- should provide the Buyer with an estimate for any additional work necessary to make a file ready for imaging.
  - If such costs are difficult to estimate, a guesstimate or a not-to-exceed amount should be provided.
  - If estimates for extra work are not provided by the Imaging Center to the Buyer in advance of the work being done, the Imaging Center proceeds at its own risk. The Buyer may choose to not pay for work done without prior approval.
  - If any work is done to the file, the Imaging Center needs to have the Buyer approve any changes.
- images the file in high-resolution
  - generates camera-ready paper output or plate-ready film output according to the Buyer's instructions. The final output should match the final low-resolution proof in content and composition.
  - guarantees that this final output is within accepted Printer specifications. The Creative Group does not pay for discrepancies inside the Print Group as long as

Many Imaging Centers require that changes or alterations must be communicated in hard copy, *not* via phone call. If time is short, faxing corrections ensures both parties of accuracy.

communication is facilitated between the Imaging Center and the Printer, and the project parameters don't change.

- proofs are made to Buyer specification, expectation, and acceptability (see Chapter 19, pages 116 and 117).

Continual breakthroughs redefine how prepress work is being done. CD-ROM (Compact Disk-Read Only) has enabled Kodak to pioneer Photo CD, a technology that has major impact on color separations.

- the storage capabilities allow many digital color images, such as photographs, to be stored and transported on a single laser disk
- five different resolutions are available: low resolution for position use in page composition, high resolution for color separation
- traditional continuous tone photographs can be converted economically to digital format
- a digital color separation is much less expensive than a conventional color separation
- less expensive separations, particularly for non-critical color applications, will enable color printing to be less costly and more available
- the quality will continue to improve.

The Production Coordinator collaborates with the Printer to

- prepare printing instructions according to press specifications.
- ensure that the film is compatible with printing tolerances.
    - If the Printer cannot utilize the film, and the Prepress Specialist did not follow instructions from the Printer, the Prepress Specialist resupplies film at no charge.
    - If the Creative Group does not ensure that the Production Coordinator receives instructions from the Printer, the Creative Group assumes responsibility for the acceptance of the film and absorbs the extra fees if the Printer cannot use the film.

VIEWPOINTS:

REPURPOSED IMAGES

"Prepress companies are becoming the link between the customer and the printer, and their role is shifting. Consider these important indicators:

- Desktop publishing is expanding rapidly throughout the advertising and publishing fields.
- The marketplace is demanding color capabilities.
- Multiplant operations are being linked through satellite transmission.

These industry trends are redefining the prepress company's role. No longer will they merely be color separators. Instead, they will be the center of the information exchange and will bring together all parties, including the customer, the printer, and the color separator."

John H. Goddard, President
Momentum Graphics
Bellevue, Washington

"How do you sell your best work in a time when most clients can't seem to afford it? The real question is, 'Can they really afford anything less?' Leaner times have forced businesses to expect results from every dollar they spend. And as we all know, their fear of the unknown in design increases with the pressure to succeed. Second-guessing can replace risk-taking very quickly.

None of us in communications, these days, can afford to produce anything but the most effective, functional, and memorable creative work. Nor can we cover up a weak idea with expensive production techniques. Smaller production budgets put more emphasis on the basic ideas being sound."

Mike Scricco, designer
Keiler & Company
Farmington, Connecticut

“ An advantage of Photo CD is the availability of multiple resolutions. We'll be using the low-resolution images in presentations tools. Then we'll be archiving the high-resolution images so we can produce high-resolution hard copy and even provide access to our images for publication. We're also testing Aldus Corp.'s Fetch and other image-location devices for research [applications]; instead of the more conventional uses, we would use the software as a public-access tool. We hope some artists may make it an original form of presentation. With Photo CD, they can tell a story, like a photo essay. ”

Edward Earle
Senior Curator
California Museum of Photography

“ Workable agreements must be made between vendor and client to establish a true partnership. We must agree on such items as:

- What services do I expect from my vendors?
- How will I pay my bills?
- What alterations will I pay for?
- When should the vendor stop work on a faulty disk?
- What approvals are necessary before starting again?
- Who is responsible for trapping?
- Who is responsible for copyrights?
- Who has ownership of the resultant data on the disk?

The list of individual items should continue until both parties are satisfied that an agreement has been met. Only through establishing a foundation to a relationship can there be profitability on both sides. ”

James Hicks
Director of Print Services, Inc.
Valentine Radford
Kansas City, Missouri

“Your success with Photo CD depends on the same thing that your success with [a scanner] depends on: your color skills with the available systems. Photo CD is raw input—at two bucks a crack. Nothing much more. As such, it can be an incredible bargain or the beginning of a do-it-yourself nightmare.

Based on my initial sense, it may be easier to overcome a slightly off Photo CD scan in Photoshop than it is to overcome the results of a scan from a crummy print. Or putting it differently, only the very best prints seem to hold the range of shadow and highlight that Photo CD so effortlessly captures. This suggests that if Photo CD eliminates anything in conventional photography, it won't be [the] film, but prints.”

Michael Lonier, "Will the *Real* Photo CD Please Stand Up?," *Pre* magazine

“Color desktop will eventually shift much (most all?) color prepress activity into the printing customer's design and production environment. Desktop publishing will radically change the way color printing is bought and, therefore, the way it must be sold. Some printers will view the changes as production problems to be avoided. Others will recognize marketing opportunities. The first opportunity (or problem) is that computers always have extensive support requirements. New users of color desktop systems will need considerable production planning and training. The printer may choose to act as a consultant, helping customers select vendors to provide research, specify systems, and begin to implement them.”

Jacques Marchand
Printing Impressions
printing consultant
Marchand Marketing
San Francisco, California

19 PROOFS

The Graphic Arts Technical Foundation (GATF) is a research, testing, publishing, and library facility for members. For information: Graphic Arts Technical Foundation, 4615 Forbes Avenue, Pittsburgh, Pennsylvania, 15213-3796, 412/621-6941.

Proofs provide the insurance that various elements of a publishing project are visually correct before the next phase begins. Various proofs verify content, typography, image placement, production preparation, color, and eventually function as a guideline for printing. The kind of proof needed depends on the phase of development, complexity of the project, the printing method, the use of color, and the approval process.

The Creative Group prepares a number of proofs for the Client's review (see Chapter 17, page 93).

- Comprehensive presentation materials (comps) show design and color intent.
- A preliminary client proof shows the formatted text and rough graphic elements in position. The Creative Group submits this to the Client for text and content refinements. This is the best time for the Client to make changes.
- A corrected client proof incorporates the changes from the Client, who checks it again for accuracy.
- A "soft" proof is reviewed on the computer monitor. Some Creative Groups show the Client images and documents on the screen for preliminary approval. Monitor color can only approximate ink on paper color. Calibration software helps to adjust the monitor to be more accurate, but the viewer of color should be careful not to expect to match monitor color exactly onto paper.
- A final low-resolution proof receives the signature of approval from the Client, and then accompanies the electronic document to the Print Group. A limitation is that low-resolution proofs may show type bolder than it will be in high-resolution. Low-resolution proofs may be made in color (from thermal wax transfer printers or color laser printers) for comping purposes and to check color specifications, but the color can be inaccurate.
- Working proofs are produced electronically. They are continuous tone and have no halftone dot structure, which means they may not truly represent the film used to make plates. A halftone dot structure is necessary to print color on a printing press. Therefore, these proofs cannot be considered prepress proofs (proofs that the Client approves

The closer proofs get to the end of the production process, the more reliable they become in approximating how the printing will look. However, the closer they approximate the actual printing, the more expensive they become to produce and correct.

that most closely approximate printing). They are mainly used during the development phases of a project to check color and registration. They are being used more frequently as a last check on color treatment before going to the expense of a film-based high-resolution proof (see below).

- Ink-jet proofs or dye-sublimation proofs, such as Iris® or Rainbow®, are made directly from the electronic file. These are often used as preliminary color proofs to show how the color specifications will look. These proofs contain no color separation halftone dot structure and show a smoother blend than a halftone dot screen can. They may sometimes look different from the printed piece, for halftone dots can coarsen the images and dyes are more vivid than printing inks.
- Digital high-end proofs, such as Digital Matchprint® or Approval®, are generated electronically—not from halftoned film. They imitate the halftone dots but may not reflect the actual screen patterns that will appear in the printed pieces. These proofs are becoming more commonplace.

Prior to printing, the Print Group uses the final high-resolution output to make prepress proofs. These approximate how the project will look when printed. This is the Client's last opportunity to give approval prior to printing. Several different proofs may be generated.

- Intermediate "random" (or "loose") proofs are made from the film. These show the Client photographic or illustrative elements only, not the pages on which they will appear. They are prepared prior to full page separations and are used to adjust color balance, later to be incorporated into the full page separations. These random proofs can be made from several of the processes listed below.
- Final prepress proofs are made from the film that will be exposed to the printing plate. These "composite" proofs incorporate all page elements together. They are high-resolution proofs, used for final Client approval. If photographs or illustrations are involved, color corrections have generally been made to these elements on random proofs. The kinds available include the following:
  - Dylux® proofs (or "blue lines") are one-color approximations made of photosensitive paper exposed to the film. These are generated for one- or two-color projects

Advertising agencies often use Veloxes® for one- and two-color advertisements. Generally, magazine publishers do not provide agencies with proofs before printing the ad. For this reason, the agency provides Veloxes as proofs for the Client. On two-color ads where one of the colors is black, a Velox provides the black image and a Color Key® or Chroma-Check® overlay shows the second color.

(the second color usually appears as a lighter image), and accompany color proofs to show format. A Dylux® proof demonstrates whether or not all the images are in register and indicates the way the pages will fold and bind.

- Velox® proofs are reproduction-quality proofs made directly from the film. They are high-resolution black images on white paper. They are accurate, inexpensive, and durable.
- Overlay proofs, such as Color Key® or ChromaCheck® proofs, show each color on a separate overlay. These proofs are an inexpensive way to check spot color, position, registration, and trapping (see Glossary). They only come in a limited number of colors, so the closest proof overlay color to the ink color is used. This proofing technique was developed for four-color process printing, but is now used more often to check two- and three-color projects. The advantages are that it is a relatively inexpensive proofing method and that individual colors can be viewed separately. The disadvantage is that the color is less accurate than other methods and may not reflect registration.
- Integral proofs, such as Cromalin® or Matchprint® proofs check process color built from the halftone dot structure of the film. They are full-page approximations of the way the printed pages will look, but are made on coated white paper (so they do not show the color on the chosen paper or how the halftone dots will behave on the paper). Currently, these are the most commonly used color proofs. Because they are made from ink-matched powders and laminated material, they have slight color limitations. However, they provide high-resolution, registration accuracy, and are reasonably priced. Experienced printers have learned how to match the reality of ink on paper to these proofs.
- Water proofs
- Signature proofs are actually printed pages made on a small special press for that purpose. The advantage is that they show the inks on the actual paper that will be used and can be printed on both sides of the sheet. The disadvantage is that they are expensive and time-consuming, and may still vary from the actual press that will be used.

- ▻ Press proofs require setting up the actual printing press that will be used and performing a short-run preliminary printing. This is the most expensive and the most time consuming proofing method, for it is test-printing the project. But it is by far the most accurate, for the project is printed on the actual paper on the chosen press. Most press proofs are printed with all the combinations of the four process colors. These "progressive proofs" allow color experts to see which colors contain problems, or how best to improve color balance. This is the only way to proof specialty inks (such as metallics or varnishes) and to predict the exact way the halftone dots will behave on the chosen paper. It makes the most sense to use this form of proofing for color- or technique-critical jobs or for very high quality jobs.

- ► The technology of color separation and proofing is advancing rapidly, and color proofs are becoming more accurate and less expensive. New digital technology is making the transition from monitor to color proof much faster and more accurate.
- ► Photo CD is allowing for inexpensive less-color critical color separations, thus bringing the cost down of OK-color printing. (See Chapter 18, page 107.)

ORIGINATING CLIENT ROLE

Advances in publishing technology compress the phases from concept to proof. Therefore, the involvement of the Client is intensified, for there are fewer stages at which to catch errors, and less time between the stages.

- ► Proofs provide the last convenient opportunity for changes before proceeding.
  - ▻ Typographic errors and editorial changes should be caught and corrected on earlier proofs.
  - ▻ Each stage of proof is provided for a specific purpose; making changes that should have been made on earlier proofs increases expense dramatically.
- ► Several proofs during the publishing process must be initialed and dated by the Client before the project proceeds to the next step:
  - ▻ final writing

MAJOR DECISIONS

- plan project
- solicit proposals
- choose Creative Group
- finalize the proposal
- select the design theme
- approve the design
- review the writing outline
- edit the first draft manuscript
- review the revised manuscript
- review the final manuscript
- approve visual components
- approve preliminary Client proof
- approve corrected Client proof
- approve final low-resolution proof
- **approve prepress proof**

- final low-resolution proof
- final prepress (high-resolution) proofs.

- Previous versions and original materials (such as photography or illustrations) with Client marked corrections should accompany the next proof for comparison and verification.
- Any corrections the Client wishes to make must be clearly marked and another proof requested if needed.
- Any preexisting mistakes discovered after each initialed proof are the responsibility of the Client.
  - The Client controls the ultimate correctness of the work.
  - The Client makes sure all sign-offs and approvals precede go-aheads.

Rarely is a proof made that doesn't lead to changes of one sort or another—and changes always increase costs. The question of who pays for the changes depends on where errors originate. If the Print Group did not understand an instruction, the Print Group absorbs the cost to make corrections. If the Creative Group needs to make an adjustment because of its own errors or changes, the Creative Group absorbs the cost. If the Client makes alterations in content, text, or other changes, the Client pays for these changes.

CREATIVE ROLE

The Art Director shares the responsibility for quality and accuracy with the Client, although the Client has the final word.

- The Art Director reviews and approves all proofs before showing them to the Client.
  - Corrections deviating from the approved design are made.
  - Any additional costs caused by the Creative Group's mistakes will be absorbed by the Creative Group.
- The Art Director describes to the Client the categories of proofs that will be used.
  - Intermediate proofs, to review color with the Client before final proofs, will help ensure Client satisfaction.
  - The Art Director can shape Client expectations with an accurate description of the proofing limitations due to budget or technological constraints. The closer the proof is to the way the project will look when printed, the better the chance that the Client will be fully satisfied with the completed project.
- The Art Director manages the status of proofs and ensures that all parties adhere to the appropriate approval processes.
  - The Art Director obtains the Client signature on proofs and never gives the "OK to Print" without this signa-

Generally, the Printer makes the prepress proofs that the Client will see. But as Imaging Centers handle more and more prepress functions, they also create more of the proofs. Generally, the Group that prepares the film for plate-preparation has the responsibility of creating proofs for the Buyer to approve.

ture. It is a way of conveying the importance of the Client responsibility and the point where the Creative Group is not liable for any errors.

- The Art Director ensures that the Designer and Page Composer also review the final proofs prior to sending them to the Printer to make sure that the printed piece will match the specifications the Designer intended.

Art Directors each have their own preferences in the judgment of what makes a good color separation. For example, some prefer a high-contrast separation while others want a more subtle effect. Some may prefer a separation to be warmer; others prefer cooler. It is the responsibility of the Art Director to communicate preferences to the Production Coordinator. The more the Art Director's preferences are understood by the Print Group, the greater the chance the first separations will be acceptable.

PRINT ROLE

Each member of the Print Group has different proofing capabilities and responsibilities, which should be outlined in estimates and quotations:

- the kind and number of proofs to be supplied.
  - Decisions of what kind of proof to generate will depend on how critical the color is to the publication. For example, a brochure on cosmetics may require much more accuracy than does a brochure on electrical equipment.
  - With technological advancements, proofs are becoming more accurate and the price is decreasing.
  - Remote proofing is becoming possible where the visual data can be transmitted to a Buyer site where the proof is generated by their color output device. This will enable suppliers to work with Clients in other locations, with multiple offices, or to transmit to several publication sources at once.
- proof limitations and variations from printed pieces.
  - There is a range of tolerances for each kind of equipment, processing, and printing.
  - There needs to be a reasonable color match between original photographs or illustrations and color separations. (Acceptability is defined by the Buyer.)
  - The Prepress Specialist is responsible for removing imperfections created by dust, scratches, fingerprints on the film, etc. Sometimes dust marks appear in the proofs. If an image to be separated has scratches or imperfections, the Buyer needs to be notified before

Paper greatly affects the way a project will look as compared to the proof. If the paper is uncoated, there will be dot gain—the halftone dots that make up the color density expand a little as the ink soaks into porous paper. This will cause the images to be muted and a little darker than the proof. On coated paper, the dots do not sink into the paper as much, and images have a crisp look. Many ink manufacturers provide swatch books that show how their ink will look on both coated and uncoated paper. With some colors, the differences are dramatic.

scanning takes place, and what costs will be entailed to fix it.

    - Ink behaves very differently on various kinds of paper. Adjustments can be made if the Imaging Center is aware of the paper to be used.
    - Any other conditions between proofing and press-room operations that may cause variances between proof and printed piece should be explained to the Buyer.

The Print Group uses the proofs as guidelines in production.

- The Print Group receives the prepress proof back from the Buyer.
    - Any corrections for the Print Group to make should be indicated on the proofs, which the Buyer returns marked "OK" or "OK with corrections," signed, and dated.
    - Anytime there is a possibility of a color shift from the approved color due to more work being performed on a file, a new proof should be generated for approval.
    - The Buyer must request any revised proofs needed.
    - The Print Group accepts any additional charges unless corrections are needed due to the Buyer's errors.
- Printing will proceed only upon the return of a marked and signed proof.
- A reasonable variation in color between proofs and the printed project constitute acceptable delivery. The reasonable variation is defined by the Buyer. If the Buyer is not available to approve a press sheet, then the best judgment of the Printer defines acceptability.

VIEWPOINTS:

PROJECT PREVIEW

“Color proofing is essential to the printing business. A color proof is an image that can be approved and reproduced with confidence. It gives the client and the production department a preview of how the job will look before it goes to press.

Since one of the functions of an off-press proof is to establish and confirm goals, the final proof should represent the finished product. For years this barrier was insurmountable. Dot gain and density couldn't be adjusted and the proofs were not made on actual stock using toners made of standard pigments. Therefore, final proofs did a poor job of simulating the press sheet.

Calling on sophisticated electrographic and microprocessor technology, Kodak broke the barrier with its Signature color proofing system. The computer controls give the system flexibility in showing variations in ink density and dot gain. The final color proofs appear on the same coated stock used to print the job and are made by toning inks that incorporate standard pigments.”

Bennett Rudomen, Manager
Electronic Printing Systems
Eastman Kodak
Rochester, New York

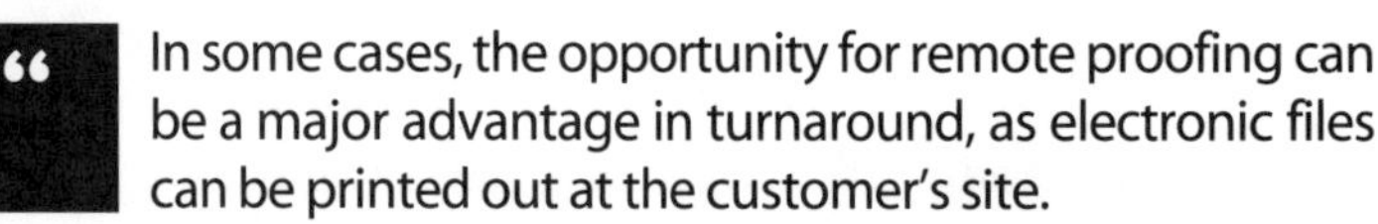

“In some cases, the opportunity for remote proofing can be a major advantage in turnaround, as electronic files can be printed out at the customer's site.

The biggest area for growth in digital proofing is opening up in scatter or intermediate proofs. Any work that requires several iterations of proofing is ripe for at least some digital work. Even if distrust in digital proofs resurfaces in the final stages, it can still make sense for all art directors and publishers to approve type, position, and color breaks, from images that show up on the desktop, especially when they are cheaper and faster.”

Steve Hannaford
journalist
Philadelphia, Pennsylvania

20 PRINTING

The Graphic Communications Association has more in-depth guidelines for printers. The GCA was organized to advance the application of new technologies in the graphic communications industry. For more information, contact the Association at GCA, 100 Daingerfield Road, Alexandria, VA 22314-2888, 703/519-8160.

Printing projects range from one-color to six or more colors, on every size and kind of paper—or even printed on surfaces other than paper. The printer should be chosen by the Buyer on the basis of their specialty and the nature of the project. Many Clients and Creative Groups work with four or five different printers due to the nature of various projects.

The printer may receive a project in one of two ways:

- camera-ready artwork provided by the Creative Group.
- film that is plate-ready, provided by the Prepress Specialist. Some printers accept electronic files directly; they have in-house prepress to output film. A small but increasing number of commercial printers are able to receive a computer disk and image it directly to press-mounted plates. When this last method becomes more widespread, it will constitute a third way to deliver a project.

Digital technology is rapidly evolving the printing industry, affecting skills, workflow, equipment, and methods.

- There are new forms of printing in development:
  - high-fidelity color (hi-fi color) that adds secondary ink colors to the cyan, magenta, yellow, and black of regular four-color offset printing. The addition of these colors increases the color gamut (the range of possible visible colors made with dots on paper).
  - stochastic screening which digitally breaks a regular dot pattern into a random dot pattern, allowing for greater color resolution in printing and eliminating moiré patterns.
- There are new technologies affecting printing equipment:
  - digital presses will enable specific quantities and very fast turn-around times.
  - high-speed customized digital presses that accept art work "direct from disk," currently similar in quality to photo copiers. Database printing, personalized publications, and small variable quantities will become more

prevalent. These presses are currently in the realm of "acceptable color" but will increase with quality.

- remote printing to off-site locations which takes the actual printing process out of the printing plant and places it at the receiver's site. This will transform some of the low-end printing of one- and two-color projects as these devices become more sophisticated (much like future high-quality color fax machines).

ORIGINATING CLIENT ROLE

The Client communicates printing needs to the Creative Group and must provide the following before printing begins:

- Quantities needed:
  - Generally, printers supply within 10% of the specified amount (see over- and under-runs, below).
  - If the Client needs a fixed amount, this must be specified.
- Approved final artwork.
- Approved proofs.
- Schedule requirements.
- Delivery instructions.
- Storage instructions, if storage of film, plates, or printed pieces are needed.

Whenever possible, the Client or Project Manager should be available

- at the beginning of the press run in case decisions need to be made.
- by phone during the press run.
- at the press run for projects of critical color to approve the color balance.

Responsibility of print quality resides with the Group who is paying the fees.

- If the Client is billed directly by the Printer, quality responsibility is shared by the Creative Group, who receives a supervision fee, and the Client, who has approval authority.
- If the Creative Group is billed by the Printer
  - the Creative Group must clearly specify to the Client and the Printer what they mean by "acceptable qual-

MAJOR DECISIONS

- plan project
- solicit proposals
- choose Creative Group
- finalize the proposal
- select the design theme
- approve the design
- review the writing outline
- edit the first draft manuscript
- review the revised manuscript
- review the final manuscript
- approve visual components
- approve preliminary Client proof
- approve corrected Client proof
- approve final low-resolution proof
- approve prepress proof
- **approve press proofs**

ity." This protects the Creative Group if their judgment differs from the Client's.

  - the Client accepts the best judgment of the Creative Group.

- Each project will have different acceptability issues. Clear communication will help satisfy both the Client and Creative Group.

The nature of the Client's needs will change in the future as technology advances.

Database technology combined with desktop publishing will enable marketing materials to be more and more personalized. This will change the way printing is ordered and used. It will also change marketing strategies. And, it will require careful accuracy to make sure that the right matches are made between recipients and subject matter.

- Color printing will become less expensive due to both cost-reductions in color separations and advances in printing presses.
- There will be greater opportunities for print-on-demand. This will allow copies of printed materials to be ordered in smaller and more exact quantities, rather than in larger quantities to be stored and used as needed.
- Printers will expand their services (see end of Chapter).

CREATIVE ROLE

Criteria for selecting a Printer:
- review samples of their work and their experience in a project similar to the one needed.
- consider the promptness of their response to the request for an estimate and completeness of their itemization.
- evaluate their understanding of the project parameters.
- ask about their capabilities for handling electronic file materials: experience in printing from electronically generated film.
- consider their presses' capabilities. For example, do they use a two-color press to print a four-color project? Do they have a six-color press for projects that require more than four-colors, or for varnish?
- compare their costs to their quality and capabilities. Are their fees competitive?

The Art Director serves as the liaison between the Client and the Print Group.

- The Art Director communicates key decisions to the Client:
  - variations in print quality tolerances.
  - variations in schedule and fees.
  - print over- or under-runs (see below).
- The Art Director supervises the printing:
  - chooses the Printer to match the appropriateness of the project.
  - ensures follow-through on Client specifications and corrections, and expectations.
  - supervises the press run.
  - manages and communicates decision making.

The Page Composer also plays an integral role in the process:

- collaborates with the Prepress Specialist to ensure the creation and output of successful electronic files. This
  - creates documents that have efficient output time.

- can save the Client and the Creative Group time and money by maximizing the most efficient use of resources.
- works directly with the Print Group during production:
  - supervises outputs.
  - receives proofs.
  - communicates with and sends all materials to the Printer.

PRINT ROLE

The Imaging Center provides

- timely and correct output according to both the Buyer's and the Printer's specifications.
- corrected materials to the Printer, notifying the Buyer of their transference.

The Printer

- reviews all proofs and materials for tolerances before undertaking the actual print run.
- provides the printed project according to the Buyer's specifications.
- provides the printed project according to the Buyer's quantities.
  - Over-runs and under-runs are not to exceed 10% (unless another percentage is agreed upon in advance) on orders for up to 10,000 copies. For orders over 10,000, the acceptable percentage should be established in advance.
  - The bill will be adjusted to reflect the actual quantity delivered. (Clients should budget for possible variance of 10%.)
  - If the Buyer requires exact guaranteed quantities, the Printer may add an extra charge in order to have enough test ("make-ready") sheets (extra sheets are needed to bring the press up to quality).
- provides the printed project according to the Buyer's quality expectations and acceptability.
  - Acceptability is defined by the Buyer during prepress communication.

Under deadline pressures, the collaboration between the Creative Group and the Print Group may be easy to overlook. Those working on the project race ahead to finish their segment and move it along fast to meet the schedule. But careful advanced planning is usually the best way to speed up the production cycle rather than trying to do the work itself as fast as possible. Lack of planning may cause more changes and delays during prepress and printing, which are very expensive and may jeopardize the deadline.

Printing Industries of America is a national association of prepress companies and printers. Printing Industries of Illinois & Indiana (PII), their largest chapter, reviewed this text and disagreed with the Client owning the final film because in traditional production, the Printer owns the film. For information: Printing Industries of America, 100 Daingerfield Road, Alexandria, Virginia, 22314. Printing Industries of Illinois & Indiana, 70 East Lake Street, Chicago, Illinois 60601, 312/704-5000.

The Association of the Graphic Arts (AGA) is a regional association of printing professionals affiliated with PIA. For information: AGA, 330 7th Avenue, New York, New York, 10001, 212/279-2100.

- ▻ The quality of the print run must be consistent throughout the run.
- ▻ Color and registration must match the approved proof, within technological tolerances.
- ▻ The finished project and the film must be delivered to the Client in acceptable condition.
- ▻ The Printer assumes full responsibility for all valuable originals while in their possession (see Chapter 28, page 160).
- ▻ The Printer is not responsible for content or for verifying the copyright of material provided to them by the Buyer (see Chapter 29, page 165).

Even if an error is made by the Printer, if it was undetected by the Client when the proofs were approved, the Client can be liable for the costs to fix the error. The Art Director needs to communicate the seriousness of the final proof approval. The Client needs to be very careful to have all necessary approvals within their own organization.

- ► is not responsible for errors that were previously undetected by the Client or Creative Group on proofs.
- ► works with the Buyer, who is often present during the press run, providing sample sheets for approval. If changes are made by the Buyer (other than to correct printing errors) that cause a loss in press time, the Buyer absorbs the extra charges.
- ► makes arrangements to store the film and deliver the printed pieces. These arrangements must be made prior to delivery (see Chapter 31, page 174).

There are new business opportunities for printers as the industry evolves and changes:

- ► increase prepress capabilities, either by expanding current systems or acquiring a prepress facility.
- ► include more fulfillment such as mailing services.
- ► offer consulting and Client on-site systems management
- ► provide training services and enable Clients to do more inhouse (which is a strong trend for Clients).
- ► further research and development capabilities (commitment to stay on the "bleeding edge").
- ► charge for storage and archiving maintenance.
- ► database management and publishing on-demand.
- ► use large memory capacity to manage Client image libraries (see Chapter 11, page 57).
- ► offer multimedia production editing and finishing operations (see Chapter 15 page 83).

VIEWPOINTS:

CUSTOM MANUFACTURING

“The free democratic society we enjoy in the United States owes its very existence to the power of the printed word. A free country must have an educated citizenry. Libraries have been called the memory of mankind. While other media can be helpful, education fundamentally depends on the printed form of communication.

Richard Weaver wrote a book called *Ideas Have Consequences,* but this is true only if those ideas are communicated. The educated citizen must be informed about public issues and controversies if he or she is to make sensible and prudent decisions. The products of printing can provide him or her with the information and ideas he or she must have to play his or her part as a free citizen in a free society.”

John R. Walter
Chairman and CEO
R.R. Donnelley & Sons
Chicago, Illinois

“Freedom of expression is guaranteed in our Constitution. Yet this guarantee gains much of its significance from the technology that allows the delivery of expression to an audience. The printed word is crucial in this process, but hardly exclusive. The digital world too bolsters this freedom, whether the medium is CD-ROM, the Internet, or future delivery systems as yet unimagined.”

Tad Crawford
Publisher and Writer
Allworth Press
New York, NY

“ An R.R. Donnelley client is making good use of on-demand services: Primis, which is part of the college division of McGraw-Hill Book Co. in New York. Professors can look through a McGraw-Hill catalog and order the journal excerpts, articles, and reviews pertinent to a particular discipline.

Donnelley pulls chapters together from its duplicate library, renumbers the pages, and generates a table of contents and index for the custom book using proprietary software developed by Kodak as part of a joint effort between Kodak and Donnelley for McGraw-Hill.

The professor then gets a bound copy that includes these materials, a table of contents, an index, and a unique ISBN number. If the professor approves the publication, the book is produced in precise quantity and delivered directly to the college bookstore for student purchase. ”

Anita Malnig and Karen Houghton, "Speedy PostScript Printers Spur Publishing-on-Demand," *MacWeek*

“ Everyone benefits from Primis. Students' costs of textbooks are kept down; the professor orders texts with chapters in the exact order of the course's syllabus, the publisher cuts down on warehouse inventory, and the author gets royalties. ”

Gerald Harper
senior systems engineer
R.R. Donnelley & Sons
Chicago, Illinois

“ Some print buyers don't know who they're printer will be as they are planning projects. Having several regular suppliers can help with this problem. Many printers are willing to help early in the project, even if they aren't awareded it later. If it's a regular relationship, as a supplier, you know you will receive another project. The client will remember that you helped them. It's important to have a longer-term view of working together than just one project. ”

Debbie Ball
printing salesperson
JB Printing
Kalamazoo, Michigan

“The future of graphic arts technology could mean the end of the highly mechanized processes now considered the bread and butter of the printing industry. The industry is on a one-way course toward the virtual extinction of press operators and craftsman as we now know them and the gain in technicians—a shift from a greater deal of physical work to more thinking.

I think the printing press of the future will include non-impact machines. The printing processes as we know them, I think, will be automated.”

Hank Apfelberg
professor of graphic communications
California Polytechnic State University

“Contrary to conventional wisdom, I don't believe the market will be dominated by a small handful of very large companies and a vast array of faceless local printers. Instead, mid-size regional printers with electronic interfaces that include advanced telecommunications will provide technical support for customers whose print inventory is maintained by their printing companies on CD-ROM and optical disks. These printers will use a variety of output equipment to provide on-demand and just-in-time reproduction services.”

Jacques Marchand
printing consultant
Marchand Marketing
San Francisco, California

"Five years ago, trade shops knew who our customers were and the services and products we provided. Sustaining our businesses required little marketing effort. Today, there are four important questions to ask:

1. Of the current products and services we provide, which will be in demand in 36 months?
2. Which of our customers will continue to need these products or services?
3. What new products and services that we don't now offer will be in demand in 36 months?
4. What new customer types may need our products and services?

These questions will help us recognize and accept the reality of change."

Paul Hanson, President
Hanson Graphics of Memphis
Memphis, Tennessee

"Some print is being replaced by other media. For example, directories or instruction manuals that often change are being handled by other media. For printers, the transition can be a barrier or an opportunity, but there are more opportunities than barriers. The barriers are set up by us, not the technology. There is a growing use for print and for new kinds of printing. And as various segments of the industry are combining, such as media buyers creating the media and media creators doing production, there are new areas for printers to explore."

Lou Laurent, Managing Director
Laurent Associates International
Sarasota, Florida

“ Desktop publishing won't be the end of the transformation of the printing industry. Why print at all? With broadband telecommunications, image and laser disks, and portable color liquid crystal displays, within a generation it probably will be cheaper to receive all this information from a hand-held terminal than it will from newsprint—information that is not only fully customized to the particular interests of each consumer but backed by libraries of supporting material for curious minds. Hints at what is to come can be found in Apple's promotion of just such a terminal, to be called the Knowledge Navigator, and Knight-Ridder's current investigations into videotex and nonprint newspapers.

The friction then for the printing industry is toward consumer choice. The consumer will be able to obtain information matched to his or her needs faster, and in the format desired. Newspapers and books will never actually go away—they are, after all, in Borges' words, 'mankind's imagination'—but the format of many will soon change from paper and print the way they once did from Sumerian clay and Egyptian papyrus scrolls.

There are several insights we can draw from the history of printing. The first is that virtual products will result from combining numerous and diverse technology advances. For example, the new printing processes are dependent upon lasers, xerography, integrated circuits, the microprocessor, high-speed communication processes, display technology, and advances in software.

The role of the author as a co-producer is also evident. He or she does not merely create the content but also can control the presentation of what is printed. To do that the author must not only be skilled in language but must also be computer literate. This is the new trade-off: greater control demands a corresponding extension in skill. ”

William H. Davidow and
Michael S. Malone,
Mohr, Davidow Ventures
Menlo Park, California

> The Various Factors Driving Prepublishing:
>
> • overlap of design and production—this does not assume the job of the creative professional
>
> • demand of information volatility—this is escalating, there is access to more information sooner
>
> • media buyers have a broader view—they see more possibilities than print
>
> • the information superhighway is under construction—more is available electronically every month
>
> • print will continue to grow—this offers the greatest flexibility is the way we present information
>
> • more flexible creativity and production techniques will be required
>
> • analog standards will be replaced with acceleration
>
> • prepress providers will expand their services
>
> • demands for database flexibility

Lou Laurent, Managing Director
Laurent Associates International
Sarasota, Florida

> The printing industry is moving from a craft-based manufacturing industry to an electronically-based service industry. Our turn-around times are shorter, the work is more customized, we will be able to only print what we need, and we will see more color content. This is not a market to push new technology, but rather is a market to pull new technology. The customers demand faster service because they are in a more competitive environment. We see a growing volume but a shrinking budget. We see a demand for more flexibility. We manufacture in hours now versus days. We need to have broader skill sets.

Charles A. Pesko
Managing Partner
Charles A. Pesko Ventures
Marshfield, Maryland

“ Two things seem indisputable: First, color desktop is coming on faster than anyone imagined possible a mere three or four years ago. Second, it will eventually shift much (most all?) color prepress activity into the printing customer's design and production environment.

If these things are true, there is a consequence for marketing: Desktop publishing will radically change the way color printing is bought and, therefore, the way it must be sold.

Some printers will view the changes as production problems to be avoided. Others will recognize marketing opportunities.

The first opportunity (or problem) is created by the fact that computers always have extensive support requirements. New users of color desktop systems will need considerable production planning and training. The printer may choose to act as a consultant helping customers select vendors to provide research, specify systems and begin to implement them. ”

Jacques Marchand
Printing Impressions
printing consultant
Marchand Marketing
San Francisco, California

“ With the advent of television and computers, people predicted the end of printing. So far, it has only grown. Computers led to an onslaught of printed materials from manuals to magazines. Television also led to special interest magazines, newspapers and books, and of course, more printed advertising. Where would America be without its Sunday television magazine or weekly edition of *TV Guide*?

For most, printed materials can be absorbed faster than spoken language. They are portable and can go wherever the reader goes. And they can be reviewed again.

Joe Kirschen, writer
Printing Impressions
Philadelphia, Pennsylvania

21 BINDERY AND FINISHING

Once the publication project is printed, the press sheets (or rolls, in the case of web printing projects) are sent to the bindery for assembly. Some printers have bindery services in-house, but others often send work to special bindery companies. Bindery and finishing operations may include

- trimming
- folding
- collating
- die cutting
- embossing
- foil stamping
- numbering
- stapling
- gluing
- laminating.

MAJOR DECISIONS

plan project
solicit proposals
choose Creative Group
finalize the proposal
select the design theme
approve the design
review the writing outline
edit the first draft manuscript
review the revised manuscript
review the final manuscript
approve visual components
approve preliminary Client proof
approve corrected Client proof
approve final low-resolution proof
approve prepress proof
approve press proofs
review bindery instructions

ORIGINATING CLIENT ROLE

Clients should be aware of how finishing work affects the schedule and quality standards of finished projects. Although the Creative Group will check advance copies, the Client should check samples from each carton delivered for consistency of quality. This should be done promptly, for if the project is incorrect, the Print Group needs to be informed within fifteen days, or they will consider the project accepted.

CREATIVE ROLE

As with other phases of the printing process, the Creative Group

- provides clear, accurate instructions to the Printer and checks format samples that serve as a guide for bindery.
- receives advance copies from the bindery to check for accuracy and quality, then gives approval to deliver the project to the Originating Client. If under strict deadline

requirements, the Creative Group may receive advance copies as the balance is sent for delivery.

- checks with the Originating Client at project completion to confirm delivery and consistency.
- approves the Printer's invoice, even when the Printer bills the Client directly. This way, the Creative Group can inform the Client of any deviations that may affect the invoice, including under- or over-runs (see Chapter 20, page 122).

Receiving advance copies helps the Creative Group to know what the Client will receive and when. Although the Creative Group sees only samples, these should represent the entire project. The Creative Group should request that the Client check samples in each carton of the delivered printed project to ensure consistency.

PRINT ROLE

The Printer generally will supervise the bindery and finishing work if done by a subcontractor. The Printer sends press sheets to the bindery as specified in the production process.

- A sample format accompanies the project to show the Binder how the finished piece is to look.
- The Printer counts the acceptable printed pieces before sending them to the bindery. Pre-counting is not the responsibility of the Binder.
- The Printer delivers printed pieces to the Binder wrapped or skidded according to the Binder's specifications.
- Containers that hold the printing remain the property of the Binder unless arrangements are made for their return to the Printer.
- The Binder is responsible for assembling the project according to the Buyer's needs for
  - accuracy—converted samples should be checked periodically for accuracy while running.
  - timeliness—deadlines are almost always a priority concern.
  - quantity—within 10% of the specified amount needed.
  - quality—cartons should be packed carefully to avoid damage of final work in delivery.
- The Binder may charge extra for preliminary samples or partial shipments of finished pieces.
- The Binder is not responsible for materials damaged due to situations beyond its control.

COMPLETION

ALTERATIONS 22

Alterations represent a change in the scope of the project parameters and specifications. Alterations made at the Client's request will likely add to the Client's costs. Changes that must be made because of the Creative or Print Groups' errors, called corrections, do not add to the Client's costs (see Chapter 23, pages 138 through 139). Alterations that cost extra include the following:

- Client requests for additional text, pages, photographs, or changes in content.
- Client requests for changes on previously approved work.
- Client change of deadline.

The Creative Group or Print Group must provide the Client with cost estimates for alterations before doing the work, or the Creative Group or Print Group is at risk to absorb these costs. When the Client knows the cost of alterations in advance, the Client may choose not to make certain changes.

ORIGINATING CLIENT ROLE

Alterations become increasingly expensive the later they are made in the development process. It is important to avoid the temptation to race ahead on a project:

- before all the pieces of a project are together (although breaking a large project into a few sizable groups of materials does help move progress forward as other portions are being completed)
- in the interest of being able to report to superiors that the project is in production (before it is really ready to be)
- before all the approvals are finished on content
- placing deadline as more important than quality unless it is.

Alterations should not be made by the Creative Group or the Print Group without providing an estimate of the costs to the Client.

MAJOR DECISIONS

- plan project
- solicit proposals
- choose Creative Group
- finalize the proposal
- select the design theme
- approve the design
- review the writing outline
- edit the first draft manuscript
- review the revised manuscript
- review the final manuscript
- approve visual components
- approve preliminary Client proof
- approve corrected Client proof
- approve final low-resolution proof
- approve prepress proof
- approve press proofs
- review bindery instructions

As a project progresses, alterations require more work, which affects cost. For example:

- a typographic change made to the manuscript may cost 20¢.
- the same change made after the corrected Client proof may cost $2.00.
- to make the change after the final low-resolution proof is approved may cost $20.00.
- on the prepress proof, it may cost $200 to change.
- when made on press, the same change may cost $2,000.

CREATIVE ROLE

The Art Director is responsible for the execution of alterations.

- The Art Director secures estimates before alterations are made and communicates these costs to the Client. The Art Director should see that the Client Group clearly understands its responsibility for paying any extra costs, or the Creative Group will be at financial risk.
- The Art Director checks for the correctness of the alterations made. (The Client, however, has ultimate responsibility for their correctness.)
- The Creative Group may need to request its own alterations due to dissatisfaction with the results on proofs. These are not charged to the Client but absorbed by the Creative Group.
- The Page Composer should be available or reachable by phone for any Prepress Specialist's questions that may arise during the execution of alterations.

As the Creative Group reviews the prepress proofs, they may be dissatisfied with the way color is balancing, the way the pages are composed, or with a technique that didn't come out as hoped. For example, perhaps the borders around photographs are too light or too dark. Extra prepress work may need to be done to correct these. Such alterations are not billed to the Client.

PRINT ROLE

Buyer may request alterations that are not included in the quoted fees. These alterations

- must be clearly marked on proofs.
- must be estimated for the Buyer before work is undertaken. If it is hard to know how much time alterations will take, a guesstimation is better than not communicating. It is best if trust has built between the Groups that results in fairness for everyone.
- will be additional charges based on prevailing rates.

VIEWPOINTS:

THE PREDICTABLE UNEXPECTED

"If you produce a winner, people aren't going to care if you're a little over budget. And if you produce a loser, the fact that you came in on or under budget is not going to help a bad situation. I mean, you've got graphic evidence, thousands of printed copies that just ain't so hot. But if you produce a magnificent piece and the budget exceeds whatever, most people won't care."

Edwin Simon, president
The Pelican Group, Inc.
formerly vice president
Sikorsky Aircraft, division of
United Technologies
Hartford, Connecticut

"If a delivery isn't absolute (and some are), it's better to be late by one day and be right than to deliver a project on time, but be wrong forever."

Patricia Maloney
Marketing Manager
Smith, Bucklin & Associates
Chicago, Illinois

"Changes become increasingly expensive as a job nears completion. At the onset of a project, management should be forewarned about the costly impact of any changes. By the same token, designers shouldn't view presswork and finishing operations as opportunities to make corrections that might have been made earlier. With a few exceptions, the earlier the designer can fix mistakes, the less costly fixing them will be, in terms of both the budget and the schedule."

Ron Coates
Boller Coates & Neu
Chicago, Illinois

“ Low-balling a job results in a chain of 'mark-downs' by suppliers who want to do business with you, as well as a chain of 'extras' and 'mark-ups' that must be added to the client's ultimate bill. This inability to adhere to the original 'low-ball' estimate naturally generates mistrust for all designers in the eyes of the client. What will become important is to just what extent these changes will alter the perception of what 'design' is—an electronic function, inexpensive, unwarranted, or a vital ingredient of corporate and product image. ”

Regina Rubino, designer
Louey/Rubino Design Group
Santa Monica, California

“ There's never enough time to do it right in the first place but always enough time to do it over when it is done wrong. ”

Anonymous

“ Clients usually say, 'Just one more change.' Always talk through client alterations with a positive attitude—put your own 'spin' on their needs and summarize alterations in written form (via electronic mail or a memo). You're not sacrificing your client's direction and your pieces carry the effective design and strong content your client demands. If you're working with a small budget, matching their personal aesthetics against actual project costs gets them to re-visit the reasons for the changes. Include an estimate of the alt cost—it's an effective reminder of the early parameters of the projects and shows you care about, and are keeping on top of, the entire project. ”

Lisa M. Bruno
project manager
Smith, Bucklin & Associates
Chicago, Illinois

CORRECTIONS 23

Corrections are changes undertaken to fix errors generated by the Creative or Print Groups, not by the Client. Changes to correct Client errors, called alterations, are covered in Chapter 22, page 134. Corrections are needed for

- typographical errors caught in the review process that were missed by the Creative Group.
- page composition inconsistencies and format errors (see Chapter 18, pages 103 and 105).
- incorrect color, position, or size of images.
- faulty prepress instructions, such as for color overlap and position.
- any other mistakes caused by the Creative or Print Groups.

ORIGINATING CLIENT ROLE

The final printed product should match the text, design, images, and proofs approved by the Client, unless the Client is otherwise notified by the Creative or Print Groups. All errors indicated by the Client should be reflected in the next step of the development process.

Within the Client organization, it is most efficient to ask decision-makers to initial proofs that they have reviewed. This is important for two reasons: it impresses upon the decision-makers the importance of their review, and it helps them to focus on finding any needed corrections as early in the process as possible (and it protects the Creative Group with there are later questions). The Client Group is ultimately responsible for the correctness of prepress proofs.

CREATIVE ROLE

The Creative Group is responsible to manage all corrections, but not responsible for the accuracy in the final prepress proof.

- During the version and approval processes, the Creative Group ensures that the final proofs reflect all Client decisions made during the process.

MAJOR DECISIONS

- plan project
- solicit proposals
- choose Creative Group
- finalize the proposal
- select the design theme
- approve the design
- review the writing outline
- edit the first draft manuscript
- review the revised manuscript
- review the final manuscript
- approve visual components
- approve preliminary Client proof
- approve corrected Client proof
- approve final low-resolution proof
- approve prepress proof
- approve press proofs
- review bindery instructions

The Creative Group is best protected from costly errors by collaborating with the Print Group in electronic file preparation to determine who is best suited to perform each task. Dangers arise in some of the page layout software color separation features. Color separation is a complicated discipline, requiring a great deal of knowledge and experience. The Designer and Page Composer should ask questions of the Print Group and not be overwhelmed by desire for control or ego.

- The Client is responsible for checking accuracy on the final prepress proof.
- The Creative Group does not charge the Client for correcting any errors caused by the Creative Group. These errors may include, in addition to the ones listed above,
  - mistakes in electronic file preparation.
  - instructions miscommunicated to the Print Group.
  - errors generated by experimental new processes. (New methods should be tested in advance).

PRINT ROLE

When the Print Group makes errors, it will provide, at its own expense

- corrected pages, images, film, or plates.
- correctly reprinted pieces.

The Print Group corrects errors, without charge to the Buyer, when the Print Group

- misunderstands or miscommunicates printing instructions.
- does not follow specifications.
- does not ask questions regarding incomplete information (when possible) and proceeds with the work.
- fails to correct any problems marked by the Buyer on the prepress proof.

The Print Group corrects errors, with a charge to the Buyer, when the Print Group receives from the Buyer

- incomplete, inaccurate, or poorly prepared copy or data from handwritten or verbally submitted text or changes.
- incorrectly prepared files.
- mismarked proofs.

SCHEDULE CHANGES 24

At the beginning of a project, the Client should set a due date for the receipt of published materials. The Creative Group generally establishes a project schedule based on working back from that due date. However, the Client may change the schedule during the project for many reasons:

- sudden market shifts that cause the audience to change.
- addition of new material.
- event date changes.
- new rulings in the law which affect business practices or product introductions.
- management reorganization.
- staffing changes.
- funding changes.
- dissatisfaction with work performed.

When the Client requests postponement or advancement of the deadline, the Creative Group and the Print Group may adjust fees and schedules accordingly.

When scheduling a project and setting the deadline, the Client needs to factor in time to approve all steps. Unavailability of the decision-makers may require flexibility in the schedule and allowances in the final deadline.

ORIGINATING CLIENT ROLE

When a Client commissions a project, they approve the proposal that covers fee payment and schedule for deadlines. Variations can occur in the schedule once work is in progress.

- If the Client puts a project on hold temporarily,
  - the schedule and deadlines need to be renegotiated with the Creative Group.
  - overtime or rush charges may be added if the Client causes delays in approval or decisions but does not change the final deadline.
- If the Client puts a project on hold indefinitely (without specifying a new deadline),
  - the Client pays for time and materials up to that date.
  - the Client may incur additional fees later for extra start-up activities, such as meetings or implementation of a new direction.

MAJOR DECISIONS

- plan project
- solicit proposals
- choose Creative Group
- finalize the proposal
- select the design theme
- approve the design
- review the writing outline
- edit the first draft manuscript
- review the revised manuscript
- review the final manuscript
- approve visual components
- approve preliminary Client proof
- approve corrected Client proof
- approve final low-resolution proof
- approve prepress proof
- approve press proofs

- If the Client cancels a project that has begun,
  - the Creative Group provides an accounting of work-to-date.
  - the Client pays for time and materials to the date of cancellation.
  - there may be a photography or illustration "kill fee" to cover lost booking opportunities (see Chapter 9, pages 50 and 51).

CREATIVE ROLE

The Creative Group accepts a project based on an agreed-upon deadline.

- If the project parameters do not change, the Creative Group
  - has the responsibility to meet the deadline.
  - must communicate to the Client, in advance, any reasons why the deadline cannot be met.
  - may not charge any extra fees to meet agreed-upon deadlines, even if over head costs increase due to work load.
- If the Client changes parameters, the Creative Group
  - agrees to the new deadline before performing the work.
  - specifies, also in advance, any additional fees that will be necessary.
- If the Client places a project on hold or cancels it, the Creative Group
  - ceases all work, compiles an accounting of work-to-date for time and materials, and invoices the Client.
  - returns or stores project materials.
  - may requote the project when it starts up again.
  - may require a cancellation fee.

Future work load is difficult, if not impossible, to predict. When a creative professional commits to a project, there is no way to anticipate exactly if more projects will also join their work flow. If really busy, the original commitment should not suffer. This balancing act is particularly challenging for freelancers. So having additional resources to call when needed can make the difference between satisfied and unsatisfied Clients.

PRINT ROLE

The Print Group accepts a project based on agreed-upon deadlines. Printers cannot anticipate exact delivery dates in advance of receiving the project to print. Many inexperienced print Buyers will promise to have a project to a printer at a certain time. A printer cannot reserve press time, for if the window is missed, overhead for the printer and downtime on the press is very expensive. A printer can schedule the project when they receive it.

Any time a project is placed on hold, and then starts up again, those working on it need to reacquaint themselves with status, reassemble the process plan, make appropriate adjustments, and proceed. This is very inefficient and can cause all Groups to lose time and money.

- If the project parameters do not change
  - but the Buyer does not adhere to agreed production schedules in providing materials, the Print Group may at that point renegotiate final delivery dates.
  - the Print Group does not incur any liability or penalty for delays beyond its direct control (such as electrical outages, floods, fires, etc.).
- If the Buyer changes parameters by canceling a project or withdrawing material before the Print Group completes the project, the Print Group
  - assumes no responsibility for the completion of the work.
  - is entitled to compensation for all work done and expenses incurred up to the date of cancellation, such as paper purchased for printing.
- If the Buyer requires overtime to meet the deadline, it is
  - quoted to the Buyer at the Print Group's prevailing rate for overtime.
  - approved by the Buyer in advance of the work performed. If this is not done, the Buyer may not have to pay fees incurred as a result of overtime.

VIEWPOINTS:

FINAL DEADLINE

"There is no guarantee that careful scheduling will prevent all problems, but it can lessen the chance of running into an error that can't be fixed. With this in mind, preparing an accurate schedule for a job should begin with the last operation—delivery. If the project is due four weeks from now, for example, the planner would begin by subtracting the days required to deliver, bind, print and separate the job, leaving, if possible, some leeway at every step.

Working backwards from the final deadline helps determine how much time the designer can profitably devote to the piece."

Boller Coates Spadaro, Ltd.
Consolidated Papers Co.
Wisconsin Rapids, Wisconsin

"By getting into prepress functions in-house, we hope to reduce costs, speed production, and simplify processes such as accounting, by reducing the number of suppliers with whom we have to work, and gain greater control over the entire production process."

Jerry Borrell
MacWorld magazine
San Francisco, California

DELIVERY AND SHIPPING

25

The Print Group agrees to send completed work to points or receivers designated by the Buyer. Many Printers deliver printed materials locally as part of their services. For long-distance delivery, special carriers are usually needed.

ORIGINATING CLIENT ROLE

Before printing begins, the Client specifies to the Creative Group, and the Print Group if the Client will be billed directly, delivery and destination instructions.

In the event of any changes in delivery instructions, fees may be renegotiated. Common changes include

- a request for advanced copies sent to the Buyer, or other partial delivery arrangements.
- a request to ship printed pieces to additional locations.

CREATIVE ROLE

The Creative Group should confirm the Client's delivery dates and locations, and communicate these to the Print Group along with the production instructions. The Buyer conveys the Client's delivery location and requirements to the Print Group, thus linking the shipper with the receiver.

The Creative Group supervises the delivery schedule, and may require advanced printed pieces to check for accuracy before the project is delivered to the Client.

PRINT ROLE

The accuracy of all shipping and delivery instructions are the responsibility of the Buyer. The Print Group

- includes delivery costs in the quotations to the Creative Group.
- is responsible for
  - safe handling of printed material.
  - delivery of materials in good condition.
  - delivery of materials in a timely fashion.
- is not responsible for loss or damage that occurs in transit beyond its control.

MAJOR DECISIONS

- plan project
- solicit proposals
- choose Creative Group
- finalize the proposal
- select the design theme
- approve the design
- review the writing outline
- edit the first draft manuscript
- review the revised manuscript
- review the final manuscript
- approve visual components
- approve preliminary Client proof
- approve corrected Client proof
- approve final low-resolution proof
- approve prepress proof
- approve press proofs
- review bindery instructions
- **check final delivery**

Many office buildings have very specific dock procedures and delivery hours. The Client Group should check for permits, permissions, or access information needed for its building and inform the dock that the project will be arriving. This will ensure that the delivery truck has access and that the Client will receive printed pieces on time.

VIEWPOINTS:

THE FINALE

“In the busy schedule surrounding a print project, delivery may seem a mundane detail. Most assuredly, it is not! The best design in the world can be eclipsed by the disappointment of inaccurate shipping. It is important to determine delivery expectations early on and keep checking on them. Having the printed materials available as promised can be the final ingredient in the success of your project—particularly with time-sensitive material. Program booklets are no good if they show up to an event late. Sometimes the timing of the delivery is more important than details on the project.”

Kris Erhart
Account Manager
Smith, Bucklin & Associates
Chicago, Illinois

“The design may be great, the proofs correct, the printing and binding superb, but if the job is delivered late or to the wrong location, all those efforts are for naught—bottom line is the printer failed. This is the black and white of being on-time—that's expected. We try to have this service throughout the process so there's no surprises at the end. I believe that the printing business is more of a service business than a product one.”

Jim Madden, president
Rider Dickerson, Inc.
Chicago, Illinois

STORAGE OF MATERIALS

26

When the project is completed, the distribution or return of project materials is determined by the owner of each (see Chapter 28, pages 157 through 160). Project materials include:

- original artwork or boards, such as illustrations or photographs.
- materials supplied by the Client.
- proofs generated by the Creative Group.
- proofs generated by the Print Group.
- digital files.
- film.

ORIGINATING CLIENT ROLE

Because the Client owns the final camera- or plate-ready materials, these should be sent to the Originating Client after printing, unless the Client arranges for storage with the Creative Group or the Print Group. The Creative Group will often store the preliminary materials and the Printer will often store the film and plates.

Clients may assume that the Creative and the Print Groups will store preliminary materials and film. However, the Creative and Print Groups may only store these materials for a limited time. The Client must specify length of storage time, requiring notification before such materials are destroyed.

The Originating Client may request that the Creative Group supervises the storage of camera- or plate-ready artwork when the project is completed.

- The Originating Client retains ownership of those specific materials only.
- The Creative Group or Print Group may charge a nominal fee for the maintenance and responsibility of storing the artwork.

MAJOR DECISIONS

- plan project
- solicit proposals
- choose Creative Group
- finalize the proposal
- select the design theme
- approve the design
- review the writing outline
- edit the first draft manuscript
- review the revised manuscript
- review the final manuscript
- approve visual components
- approve preliminary Client proof
- approve corrected Client proof
- approve final low-resolution proof
- approve prepress proof
- approve press proofs
- review bindery instructions
- check final delivery
- **return of materials**

CREATIVE ROLE

The Creative Group should retain electronic files and elements used to create final camera- or plate-ready artwork, filed by Client and project. The Creative Group keeps the files and materials for a minimum of one year after the project is printed and notifies the Client before disposal.

PRINT ROLE

Most Clients prefer that the Creative Group retains materials. In an ongoing relationship, efficiency is gained by the Creative Group resourcing all past materials. Unless a Client needs to maintain files as part of their own library (and ownership needs to be clarified), the Creative Group offers storage management as part of their continuing service.

The Buyer and the Print Group should make arrangements for the storage, return, or destruction of materials in advance. At the conclusion of the project, the Print Group

- returns all Buyer-owned materials to the Buyer, unless otherwise instructed.
- retains intermediate materials, such as scans or films, generally for one year after the printing has been delivered.
- may store Buyer-owned materials for an additional period if requested by the Buyer, and may bill the Buyer for storage.
- takes reasonable care to provide an appropriate storage environment.
- checks with the Buyer before destroying any materials, especially after the one-year time limit. If they destroy Buyer-owned materials without agreement from the Buyer, the Print Group must replace the destroyed materials at their own cost. The Print Group must notify the Buyer if the time frame for retaining materials has passed and when materials are scheduled for destruction. It can also be a way of staying in touch with current and former Clients.
- At the end of the agreed-upon storage period, the Buyer may request that materials be returned. If they do not, the Print Group should notify the Buyer if they plan to destroy the material.

VIEWPOINTS:

ANTICIPATING OBSOLESCENCE

" Storage is a billable service. Most printers now find that a lot of their floor space is taken up with tape storage. So what they're doing is turning around and saying to their clients: 'We will house this 300 megabyte file for x amount of time in case you misplace yours because we know you don't have an image management system. So that is included in the cost of the job we do for you—we'll store it for three months or six months or whatever. At that point you have to pay for the storage medium and for housing it with us.' That is an additional service a printer can offer. If a client is going to create these very large files, then they have to be responsible for them at one level or another and for some reason the printing industry has a mess on their hands because they have these back rooms loaded with racks of tapes that are unmanageable for them. "

Sandra Kinsler
PhotoLibrary Management Service
Ventura, California

" Many people are banking on the future of storage media. Database and document management is a crucial consideration. As storage media changes and evolves, there may be a new profession in the future: the data archeologist, who will need to recover data from obsolete media.

Frank Romano, professor
Rochester Institute of Technology
Rochester, New York

27 PAYMENTS AND DISPUTES

Although parties agree on fees and payments in advance, disagreements may arise either during or after a project. Terms used to describe financial considerations include:

- *payments*—compensation for work performed.
- *claims*—requests for price adjustments.
- *disputes*—disagreements over payment amounts that require negotiation.
- *liens*—the right to hold a Buyer's property until payments are made.

It is rare for publishing projects to end up in legal disputes, but it can happen. Most of these situations are caused by

- imprecise initial agreements.
- work performed without approval.
- ownership questions.
- unpaid invoices.
- broken deadline commitments.
- unclear negotiations between third parties, usually arising when the projects' decision-makers are not involved in the work or are not responsible for payment.

Clients, Creative Groups, and Print Groups may carry insurance to compensate for damage or loss. The best protection from lawsuits is prevention through clear agreements.

The Joint Ethics Committee is a source for settling disputes through mediation and arbitration procedures, their services are available to all persons engaged in the graphic communications industry. For information: The Joint Ethics Committee, Post Office Box 179, Grand Central Station, New York, New York 10163, 212/966-2492.

ORIGINATING CLIENT GROUP

Ownership transfer of commissioned materials occurs only when payment is completed. Before that point, the creator of materials owns them, and has the right to hold materials until the Client makes payment per the agreement.

With a signature, the Originating Client agrees to pay for work outlined in the final estimate or proposal. This agreement may be amended to include

- additional work that may incur additional fees.
  - When the parameters of the project change, the Client and Creative Group or Print Group need to negotiate new fees.
  - The Creative Group notifies the Client and obtains approval before performing additional work.
  - Unless otherwise agreed, the Client pays for changes in addition to the established fees when informed there will be additional fees.
- additional use for artwork beyond the project's parameters that may incur additional fees.
  - Greater use requires extra fees, which need to be negotiated (see Chapter 28, page 158 and Chapter 29, page 163).
  - The extra fees are based on Client specifications for changes to be made on work created by the Creative Group.
- outside services that may incur additional fees.
  - The Creative Group may charge a mark-up fee for supervising the work of a third-party.
  - The Creative Group may add a fee for bearing the financial obligation to the third-party. The Client may avoid this mark-up fee by having the third party bill the Client directly (see Chapter 16, page 91).
  - The Client does not own the work until the Client has paid the Creative Group *and* the Creative Group has paid any outside sources. Disputes can arise when only one party has paid another.

MAJOR DECISIONS

- plan project
- solicit proposals
- choose Creative Group
- finalize the proposal
- select the design theme
- approve the design
- review the writing outline
- edit the first draft manuscript
- review the revised manuscript
- review the final manuscript
- approve visual components
- approve preliminary Client proof
- approve corrected Client proof
- approve final low-resolution proof
- approve prepress proof
- approve press proofs
- review bindery instructions
- check final delivery
- return of materials
- **pay all invoices**

When printing is delivered to the Client, the Client should check all the boxes for quality consistency. If they store some of the boxes for later use without checking them, and they are defective, if they have not alerted the Creative or Print Groups within two weeks of delivery, the project is considered accepted. They may not be able to receive credit towards a reprint, if reprinting is necessary.

The Client will undertake and pay for all appropriate legal searches to ensure the originality of created materials.

- This protects the Client as the user of the materials.
- Legal searches are especially important for corporate identity projects.
- The Client has the right to apply for copyright on work created for its ownership (see Chapter 29, pages 162 through 165).
    - This does not extend to portions of the work copyrighted by others, such as photography and illustration, unless agreed to in writing.
    - Even if the Client does not hold the copyright for the work, the Client can be named in legal disputes with a third-party who contests the Client's right to extend use of images without compensation (see Chapter 28, page 157).

A legal search, no matter how thorough, cannot guarantee 100% nonduplication, for there are new publications that come out every day. There may also be an obscure, hard-to-find example, though it is rare that possibilities are overlooked. It is very dangerous not to search a logo or a name before using it.

CREATIVE ROLE

The Creative Group specifies compensation arrangements in the quotation or proposal.

- Terms vary from firm to firm.
    - Some firms request payment in phases.
    - Many firms bill for large projects monthly.
    - For very large projects, a retainer arrangement is often preferred.
- Additional costs should be estimated in advance.
    - The Art Director notifies the Client of additional estimated costs before doing the work.
    - The Creative Group itemizes additional fees on the next invoice.
- Ownership agreements should be part of the quotation.
    - The Art Director should communicate terms to the Client before beginning the project.
    - Ownership of reproduction materials, including camera- or plate-ready output, does not transfer to the Client until all fees have been paid.
    - If the Creative Group has performed according to terms but the Client does not make payment according to

Many graphic design firms require 1/3 in advance of the project, 1/3 upon approval of design, and 1/3 upon completion of the project.

terms, collection costs incurred by the Creative Group, including attorneys' fees and court costs, are to be paid for by the Client.

A Creative Group rarely carries insurance against the possibility of simultaneity of ideas (Chapter 8, pages 17), and therefore secures the Client's agreement to assume such responsibility. The Client, whose organization is using the published materials, is legally responsible for that use. However, it is the responsibility of the Creative Group to provide all information about ownership and use to the Client.

The Creative Group assures, to the best of its ability, that materials created are original and meet the Client's specifications.

- If the Client's parameters for original concepts are not met,
    - the Client may request further work at no cost, or
    - the Client pays for acceptable portions and expenses for work-to-date and discontinues the project.
- The Creative Group is not responsible for any writing or imagery, as approved by the Client, that may infringe upon anyone else's rights.
- If a third-party sues the Client, the Creative Group should not be party to the suit if it has provided all usage and ownership information to the Client.

PRINT ROLE

Some disputes arise when the individual Buyer who placed an order leaves their company, and the company then denies payment. The best protection against this possibility is to get a written purchase order or advance payment to secure the validity of the order.

The Buyer should pay the Print Group in full according to the agreement in the estimate.

- The Buyer company that placed the order, not it's employees, is responsible for payment on all work performed.
- The Print Group usually specifies that payment is due within thirty days of the invoice date.
- Amounts not paid when due may bear interest at the maximum prevailing state rate from the due date until paid.
- Ownership of all work remains with the Print Group until its invoices and additional charges have been paid according to agreement.
- If payment is not made within agreed terms, the Buyer is liable for collection costs incurred by the Print Group, including attorney's fees and court costs.
- All claims of defects, errors, shortages, loss, or damage of Buyer property should be made in writing within two weeks of delivery. If not, the Print Group may consider that the Buyer has accepted the work.

The importance of backups cannot be stressed enough. Everyone involved in a publication project needs to have at least two backups of all files—one of which should be off-site.

- The Print Group's liability seldom exceeds the amount to be paid by the Buyer for the work in dispute. The Print Group should carry insurance to cover inadvertent destruction of the Buyer's property. This insurance covers loss or damage to materials. Liability shall not exceed the amount recoverable from such insurance, which includes replacement of blank media, compensation for damaged or lost original artwork, or reprinting where appropriate.
- If the Print Group damages or loses original artwork, they must compensate the owner of the artwork.
  - The owner of a slide is generally paid $1,500 per lost image.
  - The owner of an illustration is paid the original creative fee.
  - The owner of lost or damaged computer files is generally not compensated by the Print Group. The creator of the files is responsible for and maintains backup files.
- The Print Group adjusts its fees if the Buyer's parameters, schedule, or quality standards are not met.

The Print Group has no responsibility in regard to the violation of any copyrights or proprietary rights.

- Because the Print Group is not involved in the creation of the content, it cannot be held responsible for invading any person's right to privacy or other personal rights.
- This indemnification includes any matter that may be libelous or scandalous, unless the Print Group created it.

OWNERSHIP

OWNERSHIP OF MATERIALS

28

Because disputes arise over ownership (perhaps not nearly as many as financial disputes, but serious none-the-less), having clear definitions is necessary. Ownership varies from project to project and should never be assumed. Rather, ownership for all publication materials should be specified at the outset of the project.

Computer technology allows an easy flow of information from one form to another, and boundaries of ownership may easily become confused if not specified in advance. Each item is subject to negotiation, and ownership agreements should be in writing. Working arrangements should specify who will own the following:

- concepts, ideas, and designs
- background material, such as video or slides
- comprehensive presentations
- images, such as
  - photographs
  - illustrations
  - transparencies
- any data entered into a computer
- preparatory materials
- computer programs
- file media, such as diskettes
- working negative and positive films
- film or paper for reproduction use
- final printed pieces.

Regardless of the agreement, ownership is not confirmed until payment is made. (see Chapter 27, pages 150 through 153).

The most difficult disputes can arise between the Client and the Creative Group at the end of the project—before printing and before the Creative Group receives payment. The Client may want to obtain electronic files for their library or other uses when they do not, in fact, own those rights. It is best to clarify such issues, in writing, at the beginning of the project.

ORIGINATING CLIENT ROLE

Many Clients rely on their Creative and Print Groups to inform them about ownership. Often, a Client organization is not in the publishing industry and may be unaware of the industry's practices. They must know what they are buying in order to avoid potential problems.

Under typical agreements, the Client owns the following (but keep in mind that everything is negotiable):

- The Client owns all material they provide to the Creative and Print Groups. This material is ultimately returned to the Client after the project is completed.
- The Client owns the Creative Group's design concepts only for the purpose specified in the proposal. Any other uses must be negotiated (see page 158).
- The Client owns final output, whether reproduction-quality paper or film (see Chapter 18, pages 100 and 101) after they have paid all invoices, even if that output is stored by the Print Group.
- The Client owns final printed pieces after invoices are paid.
- The Client owns usage rights as granted through licensing agreements (see page 158 and Chapter 29, page 163 and Chapter 30, page 169). Under these agreements, the Client may be granted the right to use:
  - concepts and ideas.
  - copies of computer files.
  - original artwork, such as an illustration or photograph.
  - final printer's proofs.

Ownership and usage will impact creative fees. For this reason, it's important that the Client fully inform the Creative Group of

- all specific uses for materials developed.
- all parameters (such as quantities, sizes, and scope), as within the project agreement.
- needed revisions or quantities that are still for the same original purpose.
- any additions.
- changes in usage.

Agreements should specify who owns final output, such as reproduction-quality paper or film, and copyright. For example, the Client may own the output but the creator may own the copyright. Copyright applies to the tangible form of a concept, not the concept itself. Computer files are covered by the copyright requirement of tangibility, because an image exists electronically and it can be printed onto paper.

If the Client does not own the copyright, the creator owns the final digital computer file. The creator is usually a member of the Creative Group. Clients who wish to obtain a copy of the file for their archives should specify this in advance. Payment for this purpose needs to be negotiated between the Client and the Creative Group. Technically, if the Client wants to alter any file in their possession but of which they do not own the copyright, permission needs to be obtained from the Creative Group. But a lot depends on the nature of the relationship between the Client and the Creative Group.

CREATIVE ROLE

While keeping in mind that other arrangements may be made, the Creative Group typically owns

- all unused concepts presented to the Client.
- all comprehensive presentation materials, including those of the approved designs (see Chapter 8, page 39 for definition).
- preliminary work:
  - original artwork
  - original computer files
  - preliminary outputs and proofs (see definition Chapter 19, page 111).
- final computer files prepared for final output (see Chapter 18, pages 100 and 101).
- the right to use printed pieces as samples for self-promotion unless the Client Stipulates otherwise.

The Creative Group is compensated by usage (versus a royalty agreement). However, book authors usually receive royalties based on the number of books sold. The Creative Group should secure written agreement from the Client on

- ownership of all materials.
- appropriate releases, such as model releases when people appear in a photograph.
- usage rights—which (see Chapter 29, page 163 for more explanation) includes:
  - *First rights*: The design or image is only used for one specific purpose, such as a brochure cover or poster. Fees do not include additional uses and the Client may not use the image without making arrangements with the creator.

Some corporations have policies that require they own all material created for them and that the creator cannot use any of the artwork for self-promotion, particularly where the Client name would be apparent.

If more applications are needed, the Client receives more benefit from that design and needs to compensate the Creative Group appropriately. When the Creative Group knows all the uses a design must address in the beginning of the project, they can design to meet that level of flexibility.

If a project meets with great enthusiasm from the Client's organization or market, the Client may wish to continue to use this design in other materials. There are additional fees from the Creative Group whenever the usage is expanded. These fees should be communicated to the Client before the theme is developed or used further. In developing good Client relations, there are skillful ways to handle usage changes.

- *Specified Rights*: The design or image may be used for additional specified purposes, if such agreement is put in writing, and the Creative Group is compensated accordingly. This can cover a series or a campaign.
- *Additional rights*: The design or image is used for more purposes beyond the original agreement, requiring additional compensation to the owner. This is often determined after the original images are created.
- *Unlimited rights*: The Client may use the design or image for any purpose without needing permission or paying additional compensation. The fee for unlimited rights is usually based on the first rights creative fee plus 50% to 100%. Some projects, such as a corporate identity or product identity, automatically imply unlimited use.

If the Client does not have unlimited rights to use a design or image, the Creative Group may negotiate fees for developing

- revisions, such as updates or changes.
- other uses, such as other forms of promotional pieces.
- series, such as additional pieces based on the same design.
- variations of the design for other purposes.
- other transformations into additional projects.

PRINT ROLE

As part of the production chain, the Print Group handles materials owned by the Originating Client Group or various members of the Creative Group. Such supplied materials include

- camera-ready art and mechanicals.
- final paper or film output that the Print Group produces.
- slides and transparencies.
- tapes, diskettes, and other digitized files and media.

In the handling of this material, the Print Group is

- responsible to receive, store, and preserve its condition but is not liable for loss and damage from causes beyond their control (see Chapter 27, page 153).
- liable for Buyer-suppled disks and other media if the media is damaged or lost. This responsibility is limited to replacement of blank media. Each organization in the Print Group should carry adequate insurance to cover possible loss or damage to materials. Liability for property does not exceed the amount recoverable from this insurance (see Chapter 27, page 153).
- responsible for return of Buyer-furnished materials after the project is finished.
- not responsible for the accuracy of Buyer-supplied materials (see Chapter 29, page 165).

The Print Group owns

- blank and recorded digital materials, including data, text, and codes that they prepare to support final output.
- intermediate hard copy and electronic materials they prepare, which may be disposed of without notice (usually within thirty days after completion of work, unless agreed otherwise).
- computer programs, systems analysis, and related documentation developed for a project. No use in whole or part can be made of these without permission and mutually agreed payment to the Print Group.

VIEWPOINTS:

POWER OF ORIGINATION

“Licensing fees are analogous to renting a car. When you rent a car, it doesn't mean you *own* the car, it means you have purchased the right to use the car. The longer you retain the car, the more money you pay.”

Jo Ann Calfee
The Stock Market
New York City

“I believe the PC revolution is part of a fundamental human desire to control your own destiny.”

G. Glenn Henry
a former IBM Fellow

“When all the rights are open to negotiation, the deal is everything.”

Richard Haukom
multimedia developer
San Francisco, California

“For most projects, creating your own is the way to go—original content frees you from licensing negotiations and fees, opens up the potentially lucrative position of licensing your content to others, and gives you the possibility of profitable spin-offs. However, producing content may be more expensive and time-consuming than you like. You must be extremely careful to have work-for-hire contracts, so your helpers don't wind up owning a slice of your CD-ROM pie.”

William Rodarmor
managing editor
California Monthly
University of California, Berkeley

COPYRIGHTS 29

Copyright protects original works of authorship fixed in a tangible medium of expression. It covers many different forms of authorship, such as the following:

- literary works
- pictorial works or graphic works
- audiovisual works
- computer programs
- sculptural works
- musical works
- dramatic works.

Currently, the duration of a copyright is for the lifetime of the creator plus 50 years. After the copyright expires, the work of the creator becomes "public domain," which means that it is no longer protected by copyright and may be used by anyone without permission.

Computers make altering and "borrowing" existing artwork easier and easier. Many who copy files or scan images are unaware of legal violation. Because it is so easy to do, it seems okay to do. But the proliferation of legal disputes indicates that this a risky practice.

The following factors determine ownership:

- The creator of a work is generally the owner of the copyright, and is wise to attach the copyright symbol, date, and name to the work to ensure protection.
- Ideas cannot be copyrighted, only their expression in a tangible form, which includes a computer file.
- If a work is created by the employee of a business within the scope of his or her employment, the employer holds the copyright.
- If a Client commissions a work from an independent contractor, the independent contractor holds the copyright—unless both parties agree in writing that the work is a "work for hire." If so, the Client owns the copyright if the work comes within one of nine categories identified in the copyright law. These are broad categories, and an attorney should be consulted if there is a question of rights. (The law is intentionally rather vague.) These categories are:
    - a contribution to a collective work (such as a magazine, newspaper, encyclopedia, or anthology).

- a contribution used as part of a motion picture or other audio-visual work.
- a supplementary work, which may include pictorial illustrations, maps, charts, etc., done to supplement a work by another author.
- a compilation (new arrangement of pre-existing works).
- a translation.
- an atlas.
- a test.
- answer materials for tests.
- an instruction text (defined as a literary, pictorial, or graphic work prepared for publication and with the purpose of use in systematic instruction activities).

Many computer data bases would be covered under collective works or compilations.

Two good sources for more information are: *The Legal Guide for the Visual Arts* (just released in a fully revised third edition) by Tad Crawford and *Business & Legal Forms for Graphic Designers* by Tad Crawford and Eva Doman Bruck. The forms are available on disk in PageMaker, both on the Mac platform and IBM platform. For information: Allworth Press, 10 East 23rd Street, Suite 400, New York, New York, 10010, 212/777-8395.

A photographer may grant the use of a photograph for a magazine within the real estate industry to be used only once. The photographer may then grant the use of the same photograph to another Client for use in ads placed in travel magazines which may extend for the life of the campaign.

The copyright owner usually maintains the original artwork, and owns it even if it is in someone else's possession. The owner has the right to

- make derivative works, including modifications.
- publicly distribute, perform, or display the work.
- control reproduction of the work—setting up licenses (see Chapter 28, page 158 for more explanation), which include:
  - *First rights* and *specified rights* are for purposes specified in the proposal (often each subsequent license carries a higher fee). These purposes may include usage rights for specified markets, media, or frequency of use.
  - *Additional rights* or *consecutive use* allows the Client the right to use a creative work in several reproductions. This is useful for campaigns and cases where a series of publications is required. Often, a bundle of rights is transferred to the Client.

- *Exclusive rights*, prohibits the creator from selling rights to anyone else. This can be limited to all markets or to a specific market. It can also be limited to a certain time period.
- *Unlimited rights* means that the Client purchases the right to use the creative work in any manner, for any market, and for either a specified length of time or an unlimited length of time. The creator still maintains the copyright and can control alterations, variations, or derivations.

All images used in reproduction must be checked for copyright by the Client prior to publication.

Permission *should* be obtained for use of an image

- in any form of copying or publication using whole or part of the image, for example:
  - photocopying
  - scanning
  - computer use
  - printing
  - photography
  - slide projection
  - presentation.
- when used as a reference by an artist or illustrator.
- when used in presentation materials, such as slide presentations or comprehensive presentation materials.
- when used in reproduction, such as printing.

Permission does *not* need to be obtained for use of an image

- when under the "fair use" clause in the copyright law, which allows uses that are not copyright infringements. Again, the law can be vague; this should be checked with an attorney.
  - These uses include criticism, comment, news reporting, teaching (including multiple copies for classroom use), scholarship, and research.

A seminar presentation for educational purposes, where examples of printed materials created by others are used to illustrate the speaker's content, would probably be considered fair use.

▷ Factors to consider when determining what is a fair use include the purpose and character of the use (whether for commercial or for nonprofit educational purposes); the nature of the copyrighted work (whether the work is published or unpublished); and the amount and substantiality of the portion used in relation to the copyrighted work as a whole.

Although it has always been the responsibility of the Creative Group to provide copyright information to their Clients, traditional processes made who owned what more clear-cut. For example, the final artwork was always a keylined board whose *usage* was granted to the Client, but copyright resided with the creator. In electronic publishing, processes are blurred, and the Client owns different materials depending on the project. There are more ways to define rights. However, uses remain the same.

ORIGINATING CLIENT ROLE

The Client and the Creative Group must decide on copyright ownership before work begins. If the Client wants to hold the copyright, the Client should

- understand the usage rights of the design and images. These rights should be outlined by the Creative Group, and the Client should be aware that fees are partially determined by these rights.
- obtain a written agreement with outside creative sources if there is need to secure the copyright.

With many projects, such as informational brochures or publications, who owns the copyright is not an issue. If the Client does not have in-house capabilities or need to own electronic files, they would work with the Creative Group on all revisions and updates anyway. An efficient system and rapport can be established which best serves the Client in an ongoing relationship.

CREATIVE ROLE

The creator (or author) holds the copyright on all work created unless other arrangements are made with the Client. The Creative Group

- bases fees partially on the uses of the design and images outlined in the proposal (see Chapter 5, page 22 for fee information.
- takes the initiative to define these uses and rights to the Client (see sidebar).
- negotiates terms with the Client.

PRINT ROLE

Copyrights generally do not apply to the Print Group.

- They are not responsible for content, but are hired to supply technical support.
- They may be called upon to suspend production during a copyright infringement violation.

VIEWPOINTS:

ULTIMATE AUTHORITY

"The test for copyright infringement is twofold: (1) Proof of access by the infringer to the work alleged to be infringed must be shown (or, if the similarity between the two works is great enough, this access can be assumed); and (2) The jury must conclude that an ordinary observer would believe one work is indeed copied from another. On the other hand, fair use has a fourfold test: (1) the purpose and character of the use, including whether or not it is for profit; (2) the character of the copyrighted work (use of an informational work is more likely to be a fair use than use of a creative work); (3) how much of the total work is used in the course of the use; and (4) what effect the use will have on the market for or value of the copyrighted work."

Tad Crawford
Allworth Press
New York City

"Thousands of copyright violations happen in corporate America every day without anyone giving it much thought: Many people mistakenly believe that copyright issues don't apply if the application isn't a commercial product. It's only that the smaller your audience and the less commercial your intent, the lesser your chance of being caught. But the risks are real, and they're growing, as content owners pay attention to the practice of casual multimedia sampling."

William Rodarmor
managing editor
California Monthly
University of California, Berkeley

“Fair use is a tricky doctrine and a trap for the unwary. People who want to use copyrighted material have devised enough rules of thumb to hold a national thumb-wrestling content. You'll hear things like, 'It's OK to take up to 30 seconds of music' and 'Two lines out of an eight-line poem is all right.' These shorthand rules are more dangerous than no rules at all. If you're producing a commercial CD-ROM and it doesn't have an explicit educational, critical, or scholarly message, this small-scale borrowing could be copyright infringement.”

William Rodarmor
managing editor
California Monthly
University of California, Berkeley

“For all works created or published for the first time after March 1, 1989, no copyright symbol is necessary for protection. That's the date the United States officially became a member of the International Copyright Treaty. By its very existence, the photo is protected by the copyright law. (For works created before March 1, 1989, and even those reprinted after that date, a copyright notice must be legible on all copies.)”

Philip Bishop, writer
San Francisco, California

OWNERSHIP OF SOFTWARE APPLICATIONS

30

Software applications include any programs purchased from software developers that are loaded into or accessed by a computer's operating system to create documents. These include the following:

- word processing programs (such as Microsoft Word® or Word Perfect®)
- drawing programs (such as Corel Draw®, Aldus FreeHand®, or Adobe Illustrator®)
- painting programs (such as MacPaint® or SuperPaint®)
- image manipulation programs (such as Adobe Photoshop®, Aldus PhotoStyler®, or ColorStudio®)
- page composition programs (such as FrameMaker®, Aldus PageMaker®, QuarkXPress®, or Ventura Publisher®)
- graphic and image resource collections (such as clip art or photo libraries)
- creativity software (such as outlining programs or authoring tools)
- digital fonts (including typeface families and image fonts)
- planning and management software (such as job tracking systems or workgroup tools)
- other business software (such as accounting packages and databases)
- utility software (including calendar programs and organizational tools).

Today, software licenses are location-specific, tied to workstations. As networking capabilities expand, documents will travel easily from workstation to workstation, following the user. This will require a new structure to licensing agreements.

Software vendors' rights and responsibilities

The purchase of an off-the-shelf software application program is actually the purchase of a licensing agreement, rather than ownership of a product. Although efforts are being made to standardize licensing agreements, vendor agreements vary. Most licensing agreements specify that:

- the software may not be copied, except for one archival backup copy.

- the program may only be used on one computer at a given time (versus being tied to an individual user).
- violators are subject to legal ramifications and withdrawal of support.
- the vendor has limited liability.

Some software companies are recognizing that a growing number of individual users own and use more than one computer. They have one at home, one at work, and a portable for in between. Licensing commonly allows a user to install on two of their own computers.

Developers have the responsibility to inform registered users of verified bugs in their software and of solutions or workarounds, if any.

Software users' rights

Although each vendor is different, most packages allow the user the following rights:

- The vendor grants the user the right to run the software on one machine at a time and provides necessary disks, manuals, and support.
- If large numbers of copies are needed, as in corporate settings, the vendor may offer "site license" agreements, where additional licenses are purchased at a discount.
  - Quantities are usually specified and generally limited.
  - Many professional associations are encouraging software vendors to provide those who only need a few extra copies with a means to purchase additional licenses without needing to buy additional disks and manuals.
  - There is as yet no standardization of licensing agreements.
- The vendor usually rescinds the right to use a version of the program if the user has purchased and installed an upgraded version. The title transfers to the upgrade, and the user may not use earlier versions on other computers, sell them, or give them away, without permission of the software vendor. The individual software licensing agreement spells out specific rights.

Piracy means copying software and giving it to someone else for use. Software piracy denies vendors income needed for further research and development of software applications.

ORIGINATING CLIENT ROLE

If the Client has requirements for how the project is to be prepared electronically (or, conversely, traditionally),

- the Client should communicate this to the Creative Group before requesting estimates or proposals.
- a particular software program should be specified in the project planning. Special purchases may be needed, additional training required, and specific design parameters developed.

- if the Client selects a Creative Group that does not use a required application, the Client should purchase it for the Group members who will need it.
- in no event should one party loan software, including fonts, to another (see Print Role following).

If the Client purchases templates from the Creative Group to be used in-house,

- the Client must purchase their own copy or copies of the software application used with or for the electronic files.
- the Client should ensure that its operating systems and applications can be integrated with the Creative Groups' systems or vice versa. Hardware configurations and software versions may vary, even within the same platform. It is best if the Client assigns an in-house technical person to work with the Creative Group to enable efficient integration.
- the Client is not responsible for the legal use of software by the Creative or Print Groups.

Desktop publishing growth is hampered by proprietary systems, both in hardware and in software. Though some packages are multi-platform, some packages can be linked together and allow some multi-tasking, the future of electronic publishing will eventually become more open. Seamlessness between packages will allow document files to utilize features from many packages. It will also require the user to learn more capabilities. Processing speed and memory capacity improvements propel more capabilities.

CREATIVE ROLE

A member of the Creative Group may need to serve as a systems integrator to coordinate with the Client's technology. The Creative Group will most often purchase, at their own cost, the programs, fonts, and other software needed for the project (unless the Client requires a specialized application). The Creative Group should

- adhere to licensing agreements.
- be adept at using the software agreed upon, and determine whether or not the software will work as intended. An exception may arise when the Client intends to use the finished computer file with a specialized application, particularly if that application falls outside common publishing packages.
- take precautions against software piracy and
  - prevent illegal copying of software.
  - prevent the borrowing of software.

- assign an appropriate individual the responsibility of ensuring proper software use. This individual informs all users about licensing and legalities, and keeps track of version numbers and upgrades.

- notify the software developer when bugs are detected that affect the efficient operating of the software according to the published capabilities. The software user (not developer) is responsible for results of the software performance, including results from the interaction between different applications.
- test new tasks, processes, and integrations before committing to project schedule and results.
- protect project components by
  - backing up all files, keeping offsite storage for archiving important files.
  - dating all versions of documents.
  - keeping track of processes relating to versions so that variations can be made if needed.

Each user can increase efficiency by developing good desktop management habits. By setting up systems for naming files, recording revisions, and backing up files, many of the most common errors can be avoided.

PRINT ROLE

The Print Group purchases, maintains, and learns its own software applications. The Print Group should also

- adhere to its own licensing agreements.
  - It must purchase its own copy of the software if the same application is used by both the Creative and Print Groups.
  - An exception is when the Print Group does not have the font required for the project. In that case, the Creative Group may provide the font along with the project files. It is included for use *only* on that project and remains owned by the Creative Group. However, some licensing agreements may prohibit this practice, though efforts are being made to standardize. When in doubt, check specific agreements.
- report software bugs to the vendor, and to Buyers, if appropriate.

As more fonts are introduced to the market, Imaging Centers cannot possibly keep up by having all fonts and versions. In order to service customers, it is becoming necessary for software companies to become more lenient about the "borrowing" of a typeface to be used only on the project for which it is intended.

VIEWPOINTS:

PUBLISHING TOOLS

“ The variety of software licensing agreements available makes it difficult for a manager to control and keep track of what type of software is under what type of agreement. Software can be licensed per machine, per user, or per server, and on a concurrent-use basis, as well as by site/entity. Often, a manager can have one of each type of licensing agreement in the office, making it difficult to keep users educated.

Our committee recommends licensing standards for a concurrent-use basis. The vendor licenses a certain number of software copies and puts them on the network. Metering software prevents users in excess of the preset number from accessing the software. The software could then be metered as part of the networking operating system.

We also request that licensing agreements be written in a more standard language that is clearly understandable to both parties. Documentation should be separated from licenses.

PC managers must educate their users on licensing issues and software piracy, as well as establish a standard policy. ”

D. Keith Heron, Co-Chairman
Micro Managers Association
White Paper on Network Licensing

“ Don't buy the first versions of a product. There will be more leasing of equipment because the technology is changing so fast. There is never enough memory. CD-ROMs will get so cheap that they can be inserted into magazines for a few cents each. More magazines will be interactive by subscription. ”

Chris Carey, teacher
Edgewater High School
Orlando, Florida

“The skills of the designer are devalued by the software developers. As designers, we should d be thinking about how much space we should delegate to each element, how elements should interact, what typeface should be used, what is the best way to relate the message to the viewer—not what 'skuzzy' port my Syquest drive is plugged into, or what file format I need to create color separations out of FileMaker Pro.

Do we complain about technology? No. We revel in it. We find that we have what we've always been missing: control. We control a project from beginning to end. We can produce design in record time. What is the price for this control? Losing design time due to learning software, debugging software, and updating software. More time is spent learning to typeset, scan photos, color correct, composite, and trap than on design. On the up side, the more we learn to do well, the more we can do in-house and bill for it. On the down side, the more we do in a shorter time, the more is expected.”

Eric White
Interactive Communication Arts Inc.
Winter Springs, Florida

“When evaluating a software package for a large installation, there are several questions to ask that buyers don't often think of: How important is the software package to the company who provides it? How much money do they have to add enhancements? Do they listen to you, the customer?”

Sandra Kinsler
PhotoLibrary Management Service
Ventura, California

“The new generation thinks nothing of copying tapes, on-line services, etc. There is no public education to teach the importance of intellectual property. There is a disrespect and even trivialization. Greed can be an overwhelming drive. There needs to be an economic incentive to inspire people to behave.”

Jerry Skapof
KanImage
Long Island City, New York

OWNERSHIP OF SOFTWARE DOCUMENTS

31

Software documents are prepared electronic files, not the software application used to create them. They are

- commissioned by the Originating Client.
- created by the Creative Group.
- refined and prepared for printing by the Print Group.

There are six characteristics of works in digital form:
1. Ease of replication
2. Ease of transmission and multiple use
3. Plasticity of digital media
4. Equivalence
5. Compactness
6. Non linearity
—Pamela Samuelson
Professor of Law
University of Pittsburgh
School of Law
Pittsburgh, Pennsylvania

ORIGINATING CLIENT ROLE

Ownership of electronic documents is not dependent on whether the Client is capable of using the electronic documents or conversant in the technology. To avoid any misunderstanding, ownership should be established before the project begins.

- If the Client hires the Creative Group to create a publication for a specific use, the Creative Group owns and retains the computer files, unless other arrangements are made (see Chapter 28, page 159 and Chapter 29, page 165)
- If the Client purchases unlimited rights usage (see Chapter 28, page 159), the Creative Group owns and retains the files.
- Only when the Client purchases the copyright can they then own, maintain, and alter the electronic files.

When competing for projects with other firms, many Creative Groups may be reluctant to bring up the issue of ownership to their prospective Clients. Ownership is a negotiating point and fees can be lower if the Creative Group retains ownership. When presenting responsibly to prospects, these concerns can be impressive—especially when the competition is neglectful.

CREATIVE ROLE

Generally, the creator of the electronic file owns the file (see Chapter 28, pages 156 through 160).

- The value of written agreements and the Art Director's role to communicate ownership should be stressed:
  - the Client needs to be clear about what they are buying and what they own.
  - creators need to specify usages and payments.

- The copyright holder also has the right to create derivative works, including copying files and altering them.
- Once a work is on computer, it is easily accessible to anyone possessing the electronic files:
  - an electronic file may not be copied without permission from the owner, unless it is in the public domain (see Chapter 29, page 162).
  - copying electronic files constitutes theft of intellectual property. No one may copy, adapt, or alter a work without the permission of the copyright owner.
  - because a file is not physically protected, professional ethics and courtesy must reign.
  - the creator of the file may protect property with a clear copyright notice alongside the image or an embedded identifying mark, such as company name or logo somewhere in the file. Whenever possible, the creator should place the work in an "encapsulated" form (such as Encapsulated PostScript) that cannot be altered.
- The Creative Group must take precautions to preserve the files created:
  - maintain two current backups of the files at all times.
  - store one backup copy offsite.

"When dealing with Clients, it is important for the artist to identify the differences between the image itself and the electronic file. If nothing is said, the Client will most likely assume he is buying the electronic version of the work as well.

"Copyright law in both the U.S. and Canada treats a work in a different medium as a different work. So, in theory at least, an artist can sell print reproduction rights only, and retain all rights in the electronic medium. When providing files for an electronic publication, the artist would give a one-time license to use the file for production purposes.

"It is best to discuss this when the artwork is commissioned and to have the details specified in a contract. Generally, when Clients understand what is at stake, they are perfectly agreeable."

— Catherine Farley, *Aldus Magazine*

PRINT ROLE

Computer documents and their component parts remain the property of the applicable Buyer, not the Print Group.

- The Print Group owns only the files they have created to enable the project to output properly. These include
  - special "print files" (see Chapter 18, pages 103 and 105)
  - codes
  - formulas (see Chapter 28, page 159).
- The Print Group is not responsible for verifying copyrighted material provided to them by a Buyer.

VIEWPOINTS:

INFORMATION FLUIDITY

“Corporate software copying amounts to illegal manufacture and distribution of someone else's product. In the last three years, the Software Publishers Association (SPA) has raided some 70 corporations. So far, only one has resulted in a company proving it was indeed not in violation of the law (perhaps not coincidentally, it was the only Fortune 500 corporation the SPA has raided).

The SPA is not indiscriminate in its raiding. It gathers solid evidence, generally, from employees (in at least four cases in New York, a temporary worker has tipped them to a company's piracy, generally because the manuals are photocopied) who may be more ethical than the management or may just be unhappy with the company and want to get back at it. Then the SPA gets a search-and-seizure award from a judge and conducts its raid.

The average information services manager seems to see the SPA as an ally in the fight to get their bosses and fellow employees to understand that copying software is a serious offense.”

Michael Fitzgerald
senior writer
Computerworld Newspaper
Framingham, Massachusetts

“Instead of making the user switch from program to program, software will let the knowledge worker of the 1990s conceive ideas, look up data in various locations, cut and paste from anywhere to anywhere, have access to central repositories—and then combine all the elements in a complete project either for output on a page description network printer or for electronic mail distribution to co-workers.”

Conrad Blickenstorfer
Chief Information Officer
Dormitory Authority
Delmar, New York

"Good business is good ethics. The technology is rapidly running away with a lot of peoples' work. There is a difference between ownership and copyright. What should you do?

1. Ask for permission.
2. Get it in writing.
3. Establish records.

Know fair use versus infringement. Part of what business is about is explaining, giving out information and material. For example, people don't *own* photographs of themselves. The biggest challenge is checking all permissions when under deadline pressure. Everyone wants to deal with companies that have good business practices because your jobs will run more smoothly."

Elaine M. Sarao
Coordinator
Bit-by-Bit
Infringement on the Arts

"If your pockets are deep enough, you can weather a law suit. Serious consequences are damage to your reputation and professional integrity."

Jerry Skapof
KanImage
Long Island City, New York

"Clients assume they own everything unless they are told otherwise—sometimes taking educational responsibility may seem contentious to prospects and clients. It is a skill to turn this to advantage in the light of good business practices. In the light of greater competition, you can make it good for the company that you are so conscientious."

Liane Sebastian, president
Synthesis Concepts, Inc.
Chicago, Illinois

"Desktop publishing exists now. The technology is out there. Copying is particularly inviting. How do you police it? In fact, the difficulties in detection are such that it is an incentive to not disseminate new works so they don't get out into general circulation. That is at odds with the very purpose of the copyright law.

The use of art in commercial settings demands distribution. It makes no sense to hoard images. Yet, if the image is distributed it will be digitized and copied often. As a creator, you just can't worry about it. *But,* when money is involved, then we all get our backs up."

Charles D. Ossola, attorney
Hunton & Williams
Washington, DC

"Multimedia really does not pose new or different intellectual property issues. Rather, the multimedia technology simply makes it easier and more desirable to copy than ever before, and thus more likely that copying will occur.

The first issue that needs to be sorted out is who owns what. Ownership issues are particularly complex because Congress has amended the Copyright Act numerous times, and which provisions apply depend on when a work was first published."

Jonathan Band, attorney
Morrison & Foerster
Washington, DC

"We've considered giving our clients a policy statement on how we treat information, copyrights, and handle confidentiality. We could have them sign a form of rights. This would inspire them to do business with a firm like ours because of our ethical concerns."

Jerry Skapof
KanImage
Long Island City, New York

APPENDIX

GLOSSARY/INDEX

Adobe System—a software company based in Mountain View, California. They are the developers of the PostScript page description language that made desktop publishing possible. They offer PostScript compatible products, such as Adobe Illustrator® (*see Illustrator*), Adobe Photoshop™ (*see Photoshop*), PostScript fonts, including Multiple Masters™ (*see Multiple Masters*).

Aldus Corporation—a software company based in Seattle, Washington. They are the pioneers of desktop publishing, with the first page composition system for the personal computer, PageMaker® (*see PageMaker*). They also offer other desktop publishing products such as Aldus FreeHand® (*see FreeHand*).

alterations—represent a change in the scope of a project's parameters and specifications. Alterations made at the Client's request will likely add to the Client's costs. Changes that must be made because of the Creative or Print Groups' errors, called corrections, do not add to the Client's costs.

applications—*see software application.*

Apple Computer, Inc.—a personal computer company based in Cupertino, California. They revolutionized publishing by pioneering desktop publishing with the release of the Macintosh computer and then the LaserWriter® in 1985. They manufacture and market computers and other hardware and software.

Approval®—high-end digital proofs, by Kodak.

artwork—materials created to develop a publication—the ingredients needed to make film and then printing plates. Artwork includes the creation of documents and finished pages, keylines, illustration, photographs, and combinations of all these elements.

background information—previous company information, history, articles on a specific industry, competitors' literature, market research data, slides and videos, and other appropriate information. This is supplied by the Client to the Creative Group as source material.

backups—an extra copy of an electronic file in case the original gets damaged or lost. Good backup habits, which include methods of saving and naming files, ensure dependability. It is best to maintain an off-site backup copy as further insurance.

bid—an estimate, quote, or proposal submitted to a Buyer. Often a bid is compared to other bids submitted by competitors for the same project.

bitmap—a computer-generated shape represented by and stored as an array of pixels. *See pixel.*

blank formats—*see dummy.*

A

B

bleed—parts of a publication that print off the edge of the page. Generally, the Page Composer leaves 1/8 inch of extended image outside the trim edges in their documents for bleed. This provides a tolerance area for trimming.

blue line proofs—one-color proofs made on photosensitive paper exposed to final film. These are generated for one-color projects, two-color projects (the second color usually appears as a lighter image), and in addition to color proofs to show format. A Dylux® proof shows if all the images are in registration, and is an indication the way the pages will fold and bind.

bugs—programming errors that undermine the efficient operating of software.

built elements—type, graphics, and photographs that are placed within a page. Some visuals are for position only, to be stripped in later by the Print Group. Others are created within the electronic document, often in layers to designate different colors. The term "built" refers to the assembling and creation of such visuals.

C

calibration—*see color calibration.*

camera-ready art—reflective artwork or keylines that are created with traditional techniques and placed in a copy camera for the Print Group to shoot film. Contains keylined instructions, text, line art, and crop marks, and in position images for reflective art.

campaign—series of related projects that support a marketing strategy. Typical campaigns may include informational brochures, advertisements, direct-mail brochures, displays, presentation materials, press kits, and proposal formats.

CEPS—Computer Electronic Publishing Systems.

ChromaCheck®—an overlay proof method by DuPont.

claims—requests for price adjustments, generally by the Buyer to their Supplier.

clip-art—designs and illustrations that come from old books or collections of previously produced work. Many are public domain images that can be used without permission.

collating—bindery operation of assembling pages in the correct sequence before finishing.

color calibration—a way to measure color so that color created on the Designer's monitor, which is based on a red, green, blue (RGB) color model, can be converted to cyan, magenta, yellow, and black (CMYK) to be printed, and have them match as closely as possible. Standards are currently being formed.

color compression—condensing large color files, especially in the case of photographs that require large amounts of memory, into smaller more portable files. After transfer they are uncompressed if more work needs to be done on them. This system is especially useful for archival storage and back-ups.

Color Key®—an overlay color proof method by 3M.

color registration—*see registration.*

color separation—full-color printing is done with four process ink colors: cyan, magenta, yellow, and black. Other colors are closely approximated by halftone screen combinations of these four colors (for example, green is a screen of blue combined with a screen of yellow). To prepare artwork in color, it must be translated into these four process colors, with one piece of film for each color. Photographs and reflective artwork need to be converted into the four process colors, which is called color separation.

ColorStudio™—a retouching and color separation software program published by Fractal for the Macintosh.

composite proofs—proofs that show how colors fit together for multi-colored projects. These proofs are used as guides for the Print Group to see how the colors register. *See also final proofs.*

comprehensive presentations—or "comps," are mockups that represent a concept to a Client. They may take the form of sketches, magic marker drawings, black & white laser mockups, color photocopies, low resolution color laser proofs, or "super comps" (that look almost printed—generally created with color rub-down type and photographic prints).

concepts—written or visual elements that convey a theme, direction, or campaign. These ideas are represented initially by presentation materials, and ultimately by finished artwork. An original concept is an idea that is inherently different from other concepts or ideas—one that has not been seen elsewhere by the creator (although all ideas are the combination of pre-existing elements put together in new ways). It is considered original if it contains no recognizable pre-existing copyrighted work.

continuous tone—color shadings where there are no halftone dot screens to define the color saturation. Continuous tone is inherent in original photographic transparencies, and appears in several proofing methods, particularly Iris®, Rainbow®, and Digital Matchprint® proofs. Offset lithography printing requires dot screens to reproduce color shadings. There are a few uncommon printing methods that will print continuous tone color.

contract proof—*see final proof.*

copy camera—a large, high-quality reproduction camera (similar to a photostat camera) where art boards are held under pressure as they are optically photographed onto film. This film is later developed, stripped, and used to expose printing plates. Camera-generated film is giving way to computer-generated film.

corrected Client proof—made by the Page Composer for the Client's review, and then refined into the final document. A corrected Client proof incorporates the changes from the Client, who checks it again for accuracy.

corrections—changes undertaken to fix errors generated by the Creative or Print Groups, not by the Client. Changes to correct Client errors are called alterations.

Corel Draw®—a drawing software application published by Corel Corporation for the IBM PC and Macintosh.

Cricket Draw®—a drawing software application published by Computer Associates.

Cromalin®—an integral proofing system developed and sold by DuPont.

crop marks —lines made by the Page Composer to indicate the trim edges of the document. These are placed in the margins outside the edges of the publication. Fold marks serve a similar function, but are dashed lines that indicate where fold are to be made.

cyan, magenta, yellow, and black—the four process colors. *See color separation.*

D

database publishing—linking of databases to page composition systems, allowing for targeted time-sensitive marketing, easily-produced directories and other uses.

design—creation of visual themes, elements, concepts, arrangements, and layouts.

desktop publishing—*see electronic publishing.*

die-cutting—special bindery operation for unusual cuts, such as windows cut in covers of publications, pockets for the back of folders, shaped projects such as labels, unusual envelope sizes, or inserts. Dies are made from metal razor-edged rules set into a wooden holder. A special press holds this mechanism, which stamps the paper much like a cookie-cutter.

digital high-end proofs—electronically generated proofs such as Digital Matchprint® or Approval®. They are not made from halftoned film but imitate the halftone dot structure, and may not reflect the actual screen patterns that will appear in the printed piece. These proofs are becoming more commonplace.

Digital Matchprint®—digital high-end proofs by 3M.

digital presses—new printing presses that accept digital information directly from disk to make plates. *See direct-to-press printing.*

digital scanning—capture of photographs or line art into digital form that can be used in a computer, stored on magnetic media, etc. The scanner is a device that "reads" images and converts them into digital information. These files may then be incorporated into electronic pages or output separately.

direct-to-press printing—as electronic publishing advances, digital presses are able to accept information directly from computer disk to create plates on press (without film and stripping).

disputes—disagreements over payment amounts, which require negotiation, generally between a Buyer and a Vendor.

documents—*see software documents.*

dot gain —effect seen with printing on uncoated, porous paper, where the halftone dots that make up color density expand as the ink soaks in. This can happen to a lesser degree on coated paper. Because the dots actually print larger than shown on the film, the printing comes out darker than intended—unless dot gain is compensated for in prepress preparation.

download—transferring information from one hardware device to another. Typically, special typefaces can be downloaded from a computer to a printer, or a file is downloaded from one computer to another via a modem.

dpi—short for dots-per-inch, a measure of printer or monitor resolution.

drawing programs—illustration software applications that utilize vector graphics (unlike paint programs, which are based on bitmapped graphics), such as FreeHand® or Corel Draw®.

dummy —a blank format or mockup to show paper stock and weight, size, and bindery. A dummy is usually created by the Print Group to provide the Buyer with a "feel" of the project, to function as a production guide, and to form a printing budget. Sometimes used by the Creative Group to sketch in the arrangement of content and components.

dye-sublimation proofs—continuous tone proofs, such as Iris® or Rainbow®, made directly from an electronic file. These are often used as preliminary color proofs for color specifications. They contain no dot structure and show a smoother blend than a halftone dot screen can. They may sometimes look better than the printed piece, for halftone dots can coarsen the images, and dyes are more vivid than printing inks.

Dylux® proofs—*see blue line proofs.*

electronic publishing—publishing done on computers. Originally called desktop publishing, but now includes midrange and high-end systems. One term is used because the various systems are blending. Electronic methods streamline and combine many traditional processes. Writers work on word processors, so text is entered only once. The word processing file is imported into the page composition system, where the Page Composer creates the documents. All traditional keylining is handled electronically. Some color separations can be handled on the desktop, some by higher-end electronic systems. Film is generated electronically, and in some cases plates can be prepared electronically.

embossing—special bindery operation similar to die-cutting. A special press holds a custom die to impress images onto paper to achieve a raised surface. Common uses of embossing are raised logos, lettering, or illustrations on stationery, book covers, letters on plastic credit cards and badges, raised greeting cards, etc.; and other specialty uses.

emulsion—the light-sensitive coating on one side of a piece of film or paper.

EPS files—Encapsulated PostScript files contain all the relevant information for importing images into other software packages. Generally, they cannot be altered. This is a common file transfer format. *See TIFF and PICT.*

fatal error—when an electronic document will not open for the Imaging Center.

film—transparent material that is coated with photosensitive material, or emulsion. Exposed film becomes black after development, while unexposed areas become transparent. Generally, the transparent areas will print; this form is called film negatives. Film positives are when the black areas will print.

final low resolution proof—the document prepared by the Page Composer, on desktop printer, as the last step before outputting the file for printing. The Client needs to check all the elements and proofreading. Beyond this point, changes may entail extra time and fees for the Client. The final low resolution proof accompanies the electronic document to the Print Group. Low resolution proofs may be made in color (from desktop printers) for comping purposes or to check color specifications, but the color can be inaccurate. *See low resolution proofs.*

final proofs—are made from the film that will be exposed to the printing plate. These composite proofs incorporate all page elements together. They are high resolution (or contract) proofs, used for final Client approval. If photographs or illustrations are involved, color corrections have generally been made to these elements on random (or loose) proofs.

E

F

finished artwork—*see artwork.*

folding—bindery operation that folds pages as specified in the prepress format instructions. Some projects need to be scored before they are folded.

foil stamping—similar to the die-cutting or embossing processes. A special press holds a wide ribbon of foil material, which is applied to paper when a heated die strikes the foil. The foil that was not heated and pressed onto the paper falls away. Foils are generally shiny, like silver or gold, but are also available in a wide range of colors and surfaces.

fonts—a single typeface (which includes all letterforms in the alphabet, numbers, and punctuation). In traditional typesetting, each font was a single size, and had to be formed individually. With PostScript fonts, a typeface drawn in one-size and digitally can be scaled to any size.

formats, blank—*see dummy.*

formats, typographical—a group of type specifications such as size, typeface, style, and leading (line space). Often style sheets are set up so that text can be converted into its typeset (or formatted) form efficiently. *See style sheets.*

four-color process printing—printing with the combination of the four color process inks: cyan, magenta, yellow, and black. *See color separations.*

FPO—for position only. Indicates an item intended to be replaced. *See built elements.*

FrameMaker®—a page composition software application published by Frame Technology, based in Palo Alto, California, for the NeXT, Macintosh, PC Windows, and UNIX workstations.

FreeHand®—a PostScript drawing software application published by Aldus Corporation for the Macintosh and IBM PC.

G

gluing—bindery process using adhesive for special publication needs. The spines of flat-edged books are generally glued as the cover is placed on. The pockets in the back of folders are usually glued. Some gluing, as in the latter, is accomplished manually.

H

halftone dots—individual element of a halftone screen that reproduces color shadings to simulate continuous tone photographs or illustrations. Screens are available in many resolutions, where the size of the dots and the number of dots-per-inch vary. Digitally, halftone dots are composed of pixels. *Also see screens.*

hard copy—print outs onto paper from digital files, used as a reference for checking documents.

high-resolution—generally over 1,000 dots-per-inch, high enough for output to appear perfect. For halftones to appear perfect, they need to generally be imaged at over 2,000 dpi.

IBM—International Business Machines, a hardware company located in Armonk, New York that manufactures computers.

Illustrator®—a drawing software application published by Adobe Systems for the Macintosh, IBM PC, and NeXT.

imagesetter—a device that produces the high resolution output from electronic documents onto paper or film.

imposition—the arrangement of pages prepared for a press sheet so that after printing, the pages fold to be sequential. "Printer's spreads" are this arrangement of pages so that the four-page signatures when folded are sequential. "Reader's spreads" are used in flat presentations, where the pages will not be folded or trimmed, to be sequential.

ink-jet proofs—are made directly from the electronic file with colored ink sprayed onto paper. These are often used as preliminary color proofs to show how the color will look.

integral proofs—proofs that check process color built from the halftone dot structure of the film. They are full-page approximations of the way the printed pages will look, but are made on coated white paper (so they do not show color on the chosen paper or how the halftone dots will behave on the paper). Currently, these are the most commonly used color proofs. Because they are made from ink-matched powders and laminated material, they have slight color limitations. However, they provide high resolution, registration accuracy, and are reasonably priced.

interactive media—a non-printed form of publishing where the file is opened on the computer of the recipient. The creators of interactive documents program the possible choices; the recipient can immediately access the information that interests them most.

integrated publishing—the combination of an interactive presentation with a printed brochure. Many integrated publishing projects have the disk included in a pocket of the printed brochure.

intermediate proofs—single-image proofs made from high resolution film. More colloquially called "random" (or "loose") proofs, these show the Client the photographic or illustrative elements separately from their pages. They are prepared prior to full page separations and are used to adjust color balance, later to be incorporated into the full page separations.

interplatform exchange—a method where a document can be created in one software application and then opened in another, to utilize the second application's features. Editable PostScript, when released, will make this possible.

Iris® **proofs**—are ink-jet continuous tone color proofs by Scitex.

keyline—manual composition of elements resulting in a pasteup. This artwork is often called a keyline, but the term keyline is a coding method of guides and marks carrying printing instructions for where color or images are to be inserted. In electronic publishing, such keylines are handled by the computer and the term Keyliner (the person who creates keylined artwork) is now the Page Composer.

I

J

K

L

LAN—short for Local Area Network. Used to link workstations in one location and requires a dedicated computer as a server. Collaborative computing as "groupware" is becoming more prevalent: Documents can be worked on by more than one user at a time and resources shared.

laminating—special bindery process of applying clear plastic film material onto printed pieces. A heat or adhesive process bonds the material to the paper. It gives the printed piece a highly glossy appearance, and is often used as protection on menus or placemats, manuals, etc.

lien—the right to hold a Buyer's property until payment is made.

line art—artwork that has no halftone dot or shading. It is black or white and requires no replacement screens in prepress.

Linotronic® output—a high resolution imagesetter manufactured by Linotype-Hell. The term is often used generically—and incorrectly— to refer to high resolution output from any imagesetter.

lithography—the most common kind of printing in use today.

live file—opening a document file in the original software application to make a "print file."

local area network—*see LAN.*

loose proofs—*see intermediate proofs.*

low resolution proofs—generally, proofs under 1,000 dots-per-inch.

lpi—short for lines-per-inch, a measure of halftone screen resolution. *See dpi.*

Lumena®—color painting and retouching software application published by Time Arts for the IBM PC.

M

Macintosh computer—hardware workstation manufactured by Apple Computer. The Macintosh with the Apple LaswerWriter and early software such as Aldus PageMaker, pioneered desktop publishing.

MacDraw®—drawing software program for the Macintosh computer, published by Claris Corp.

MacPaint®—paint software program for the Macintosh computer for low resolution images, published by Claris Corp.

Matchprint®—an integral proof product by 3M.

mark-up policies—a handling fee or percentage added onto the cost of outside purchases. This is to cover the financing, accounting, debt carrying, and management of the project.

mechanicals—artwork that possesses all the assembled elements and instructions for the Printer to shoot film. *See keyline and camera-ready art.*

Microsoft Corporation—a Redmond, Washington based software company that develops both operating systems (such as MS-DOS® and Windows® for the IBM) and applications (such as Word®, Excel®, Project®, etc.).

Microsoft Windows—*see Windows.*

modem—a device for using the telephone to link two computers. The computers can be different kinds and still exchange files and other information. Telecommunications software is needed to make a modem usable. Modems are the basis for calling up on-line information services such as CompuServe®.

multimedia—*see interactive media.*

Multiple Masters™—font technology originated by Adobe Systems. A typeface that can be altered by the user to be wider or narrower, thicker or thinner, or other variable characteristics. (This book is set in Multiple Masters Myriad).

network—computers that are connected for the sharing of information and resources. Often involves an extra computer, called a file server, to store and distribute files to all the other computers on the network. *See LAN and WAN.*

NeXT —hardware workstation, based on the UNIX operating system, manufactured by NeXT Computer in Redwood City, California.

nondisclosure agreement—a document generated by the Client that specifies information given to the Creative Group and the Print Group are confidential and cannot be shared with anyone outside their organizations. The agreement binds the entire Creative and Print Groups to confidentiality concerning the project.

notebook computer—a battery-powered personal computer with a flat screen monitor that folds for travel, allowing for a light-weight and great portability.

numbering—special printing process where sequential numbers are imprinted on each impression. For example, many forms need individual numbering. This process is used in specialty direct mail campaigns, forms, etc.

N

off-the-shelf software—application software packages as purchased from a vendor.

one-color printing—a single ink color applied to paper. It can be black or a special ink color. The color of the paper is not considered a color in printing.

operating system—the foundational computer software that supports software applications. The operating system of an IBM PC is Disk Operating System (DOS or MS-DOS) or OS/2, or Windows. For the Apple Macintosh it is System 6 or System 7. For the Sun or Apollo Workstation, it is UNIX. The operating system is the most basic software that makes the computer usable, ruling basic functions as saving, copying, printing, etc.

original concept— *see concept.*

out-of-pocket expenses—materials, messenger services, etc., purchased outside the organization. Common out-of-pocket expenses include mailing costs, paper, film, stats, cabs, telephone calls, etc.

outside purchases—services or materials that an organization does not handle in-house. For a design firm, outside purchases could include photography, illustration, writing, prepress, etc. If the organization is billed for services that are then billed to the Client, the service provider is a "third-party" firm. The purchaser becomes a liaison between the Client and the service provider.

overlay proofs —show each color on a separate overlay, such as Color Key® or ChromaCheck® proofs. Such proofs are an inexpensive way to check spot color, registration, and trapping. They only come in a limited number of colors. This proof technique was developed for four-color process printing, but now is used more often to check two- and three-color projects.

over-runs—variations in the quantity delivered (from quantity ordered) from a Printer. Over- and under-runs are generally not to exceed 10% on printing orders for up to 10,000 copies. For orders over 10,000, the acceptable percentage should be established in advance.

O

P

page composition—creation of the electronic documents that contain all the text and graphic elements assembled into page layouts. A number of intermediate proofs can be created that provide the Client and the Creative Group flexibility in the development of final documents—without the time and costs associated with traditional methods.

PageMaker®—a page composition software application published by Aldus Corporation for the Macintosh and IBM PC. Modules for seamlessly adding special capabilities to PageMaker are called Aldus Additions™.

paint programs—bit-mapped software applications to create pixel-based images, such as MacPaint® or PixelPaint®.

Pantone® color—a widely used color system that supports controlled ink colors. Many page composition systems have keys into the Pantone numbering system. When using process printing, Pantone colors can only be approximated.

parameters—the specifications of a project that include goals, audience, scope, size, deadline, etc..

pasteup —putting adhesive on traditional typeset galleys and photostats, then applying them to keylined boards. Doing pasteup means the same thing as keylining.

payment—compensation for work performed.

pen-based input—until recently, the keyboard, mouse, or scanner were the primary input methods. Certain new computers do not require the user to be able to type, but rather to handwrite and select options using a pen.

Photoshop®—an image scanning and retouching software application published by Adobe Systems.

piracy—*see software piracy.*

PICT —Macintosh software format for storing and transferring visual information from drawing or painting programs.

pixel—short for picture element, a single dot on a computer screen. *See bitmap.*

plate—a printing plate made of a photosensitive coating applied to thin sheet metal. Stripped film is used to expose ("burn") the plate to light. After developing the plate, the areas where there is image attract ink, while areas that should not be printed resist ink. When the plate is mounted on press, the inked areas transfer to the paper.

platform—hardware or software setup. For example, a Macintosh computer and an IBM computer are different hardware platforms. Aldus PageMaker® and QuarkXPress are different software platforms.

portfolio—samples of a Creative Group's work, usually as finished printed pieces done for other Clients. These are shown to demonstrate the Designer's experience and approach to design.

PostScript® —a proportionally-based programming language for describing images in terms of printing onto paper. Developed by Adobe Systems in the early '80s, it has become the industry standard device-independent page description language.

preliminary Client proof—a low resolution proof showing the rough formatted text and rough graphic elements in position. The Creative Group submits this to the Client for content and style refinements. This is the best time for the Client to make changes.

prepress—preparation of electronic files for printing by adding all the necessary technical specifications and outputting film. After the electronic documents have been proofed and approved in low resolution, they go to an Imaging Center for prepress functions. The developed film is then stripped into position (or electronically placed imposition) used to "burn" the plates that are mounted onto the printing press.

prepress proofs—proofs the Client approves that most closely approximate printing. Final prepress proofs are made from the film that will be exposed to the printing plate. These "composite" proofs incorporate all page elements together. They are high-resolution (or contract) proofs, used for final Client approval. If photographs or illustrations are involved, color corrections have generally been made in these elements on random proofs.

press proofs—require setting up an actual printing press to do a small run preliminary full-run printing. This test-printing the project is the most accurate and expensive proofing method. This is the only way to proof specialty inks (such as metallics or varnishes) and to predict the exact way halftone dots will behave on the chosen paper. *See progressive proofs.*

press sheets—a large sheet of paper for a printing press that may accommodate many pages of a printed document. Most printing projects up to a quantity of roughly 50,000 impressions are printed "sheet fed." This means that the images are printed on large sheets of paper, assembled, and then trimmed down. For larger quantity projects, they are generally printed "web fed" or on printing presses that handle large rolls of paper.

presentation materials and comps—*see comprehensive presentations.*

print file—a file of PostScript code saved to disk and ready to be sent straight to the printer or imagesetter. Advantages of using print files rather than printing the document from the application (*see live files*) include a cleaner trace if anything goes wrong; inclusion of fonts in the document; ability to link files together for continuous imaging, etc.

print-on-demand—allows copies of printed materials to be ordered in smaller and more exact quantities, rather than in larger quantities to be stored and used as needed.

proofs—approximations of printing at various stages before printing occurs. The earliest proofs are typically low resolution galleys of typesetting. The latest proofs are typically composite proofs made from the same film that will expose the printing plate(s).

proofread—checking over text for any errors.

process color—color printing using the four process colors of cyan, magenta, yellow, and black. Most colors can be achieved through combinations of halftone screens of these form colors. They can produce the widest range of colors with the fewest number of inks.

progressive proofs—proofing each color of a project separately and in various combinations with the other process colors. For example, the blue image is proofed alone, then in combination with the yellow, then in combination with the yellow, and magenta, etc., until all fifteen combinations have been printed. This is generally done with press proofs. *See press proofs.*

Q

QuarkXPress®—a page composition software application published by Quark, Inc. based in Denver, Colorado for the Macintosh and PC Windows. Modular additions are called QuarkXTensions®.

QuickDraw®—is a screen representation model within the Macintosh operating system. It is also used to print to dot-matrix and other non-PostScript printers, by duplicating (perhaps in higher resolution) what shows on the screen. Most Macintosh software generates PostScript for printing to PostScript printers and imagesetters.

R

Rainbow™—dye-sublimation continuous tone desktop color proofer from 3M.

random proofs—*see intermediate proofs.*

reflective art—photographs, illustrations, or other two-dimensional visuals that are created with traditional means such as prints, paintings, drawings, etc. In traditional production, these need to be mounted into a copy camera for color separation. With the introduction of drum scanners, it became more efficient to shoot a photographic transparency of reflective art, to mount around the scanner's cylinder, for color separation. With electronic publishing, less reflective art is used, for more artwork is created digitally.

register marks—a symbol (usually a cross and circle with the same center) placed outside the trim of a publication, that guides color alignment. There are register marks that the Page Composer places on electronic documents, other register marks used in prepress, and still others that appear on press sheets. Sometimes the same register marks are used throughout all three phases.

registration—how one color image fits with another color image when printed. Because images of different colors are on different plates in the printing press, how they print in relation to one another is critical to print quality. Generally, out-of-register printing leaves white gaps between colors and seems to blur color halftones.

research—includes reading previous publications, investigating sources for information, gathering information, collecting factual data, and interviews with appropriate people, such as Client management staff, customers, and suppliers.

Researcher—person who finds all relevant facts, handles any market research, accesses resources, etc..

RIP—a Raster Image Processor interprets an electronic document's digital information, converting it to dots for high resolution output.

rush charges—charges resulting from overtime or priority work with the Print Group. The Group requesting the changes is normally responsible for such charges. Generally, the Creative Group does not have rush charges but is compensated for Client changes through a direct estimate of those changes.

S

scanning—*see digital scanning.* Also refers to a traditional color separation process where images are placed in a camera or on an analog scanner.

screens—color or shading created by halftone dots. Screens are measured by the shape of dots and number of dots-per-inch. The dots have space between them which gives the effect of a lighter color. The lower the screen percentage, the lighter the color.

separated proof pages—final low resolution proofs that have one page (printed in black) for each color layer. This shows the Print Group how the color separations should come out.

sheet-fed printing—a common printing method that uses sheets, rather than rolls of paper. *See web printing.*

signature proofs—actual printed pages made on a small press especially for that purpose. They show the inks on the actual paper that will be used for printing, and can be printed on both sides of the sheet.

soft proof—image reviewed on the computer monitor. Some Creative Groups show images to the Client on the screen for preliminary approval.

software application—programs purchased from software developers that are loaded into or accessed by a computer's operating system to create documents.

software document—a prepared electronic file, as opposed to the software application used to create it. A document is generally commissioned by the Originating Client, created by the Creative Group, and printed by the Print Group.

software piracy—the borrowing or copying of software. Because this is so easy to do, many users do not realize that this practice is illegal. It is also very dangerous for the development of the software industry, as income needed for research and development of future products is lost. If a user wishes to try out a software package, they are best advised to go where it is being demonstrated or sold to try it out.

spot color—refers to how color is handled without color separations in many two- or three-color projects. The printing inks can be specialized colors. The second or third color is handled in selected areas such as text or line elements. Many page compositions are quite sophisticated in their handling of spot color separations.

stapling—a bindery operation of fastening a publication together using a large stapler. For booklets and publications with staples along the spine, the term is called saddle-stitching.

stat camera—a large copy camera that is configured to shoot line art to size for keylining into camera-ready artwork. Stat is short for photostat.

strip-ins—last minute-changes made on high resolution paper output. Just the text or images that change are pasted onto the boards.

stripping—assembly of high resolution film into the configuration, or imposition, necessary for exposing plates. In electronic publishing, pages are often composed in position.

stock images—images that already exist, either as outtakes from previous creative projects, or made specifically for sale as stock. Stock images may exist in traditional or digital form.

style sheets—*see formats, typographical.*

SuperPaint®—a paint program for the Macintosh, published by Aldus Corporation.

system configurations—the hardware and software comprising a computer system.

T

templates—electronic page grid and style sheets for design and type that is the format into which text and visual elements are to be inserted. Templates are set up for recurring publications such as newsletters, magazines, brochure series, reports, or proposals. Many Clients hire design firms to create templates which are then carried out by the Client's in-house staff from issue to issue.

third-party services—*see outside purchases.*

TIFF (Tagged Image File Format)—a set of standards for grayscale or color images, usually scanned, for transfer, storage, or placement into an electronic document.

traditional publishing—began the same way as the electronic process: with a Client-inspired project. The manuscript was typed on a typewriter, then re-keystroked by a typesetter into a typesetting machine. Rough galleys (repro proofs) were generated for proofreading, followed by corrected final galleys. The production artist (keyliner) trimmed the final galleys, applied adhesive, and keylined them onto boards. Illustration and photography were handled by "position stats" (black and white presentations), also applied to the boards. The Printer received the keylines, with the illustrations and photographs separately, to place into a copy camera to shoot film. Color images were color separated on a copy camera using filters, or, more recently, placed in a scanner. The resulting pieces of film were stripped together into flats (large sheets that put all film in position to match the way the pages will print on the press sheet). The film was contacted (exposed by light in a vacuum frame) to a photo-sensitive lithographic printing plate. The plate was mounted onto the printing press where it accepted the ink, transferring it onto the paper. (The last several steps are still done the same way with the new electronic process, but are giving way to digital means just as the earlier steps did.)

trimming—a bindery operation which takes the assembled press sheets and cuts the publication to the specified size. *See bleed.*

trapping—when one color overlaps another so there is no white space showing when they print slightly out of register. This overlap is traditionally handled by the Print Group, but with the advances of electronic publishing, more and more Page Composers are handling it in their page composition system.

TrueType™—a recent typeface system created jointly by Microsoft and Apple.

U

under-runs—*see over-runs.*

UNIX®—a multi-tasking operating system used in Apollo, Sun, NeXT, and other workstations.

vector graphics—a kind of graphics (such as PostScript) characterized by direction and scale pixels rather than bits. *See bitmap.*

Velox® proofs—reproduction-quality proofs made directly from film negatives. They are high-resolution black images on white paper.

Ventura Publisher®—a page composition software application published by Xerox Corporation for the IBM PC and Macintosh.

WAN—Wide Area Networking, a hard-wired system similar to a LAN but designed to operate efficiently over greater distances. *See LAN.*

web printing—a printing method that uses rolls (webs) rather than sheets of paper, to achieve great printing quantity and speed. *See sheet-fed printing.*

windows—rectangular panels on the computer screen for showing messages, documents, options, etc. The Macintosh was the first mainstream computer to use windows, which were then mirrored for the IBM PC by Microsoft in their operating system called Microsoft Windows.

work-arounds—alternate method or short-cut for achieving a result (usually with software) that can't be directly achieved.

working proofs—an electronically generated intermediate proof, generally with no halftone dot-structure. They are mainly used during the development phases of a project and are not intended to be seen by the Client.

V

W

INDEX OF QUOTED SOURCES

Special thanks to all those who cooperated by lending their voices and experiences to this book:

ComputerWorld newspaper, Framington, Massachusetts
Graphic Arts Show Company, sponsors of the Conceppts conference
Graphic Design USA, Kaye Publishing, NYC
MacWeek, San Francisco
Ellen Shapiro, author of *Clients and Designers*, © 1989 by Ellen Shapiro, Watson-Guptill Publications, NY
John Van, © Copyrighted Chicago Tribune Company.

INDEX OF ORGANIZATIONS

Advertising Photographers of America (APA)
1001 W. Adams Street, Chicago, IL 60607
312/455-8899

American Center for Design (ACD)
233 East Ontario, Suite 500, Chicago, IL 60611
312/787-2018

American Institute of Graphic Arts (AIGA)
National Office, 164 Fifth Avenue, New York, NY 10010
212/807-1990

American Society of Composers, Authors, and Publishers (ASCAP)
1 Lincoln Plaza, New York, NY 10023
212/595-3050

Association for the Development of Electronic Publishing Technique (ADEPT)
2722 Merrilee Drive, Fairfax, VA 22031
703/698-9600

Association of the Graphic Arts (AGA)
330 Seventh Avenue, 9th Floor, New York, NY 10001
212/279-2100

American Society of Media Photographers (ASMP)
14 Washington Road, Suite 502, Princeton Junction, NJ 08550
609/799-8300

Association of Imaging Service Bureaus (AISB)
5601 Roanne Way, Suite 608, Greensboro, North Carolina 27409
800/844-2472

Graphic Artists Guild (GAG)
National Office, 11 West 20th Street, 8th Floor, New York, NY 10011
212/463-7730

Graphic Arts Technical Foundation (GATF)
4615 Forbes Avenue, Pittsburgh, PA 15213-3796
412/621-6941

Graphic Communications Association (GCA)
100 Daingerfield Road, Alexandria, VA 22314-2888
703/519-8160

Independent Writers of Chicago (IWOC)
7855 Gross Point Road, Unit M, Skokie, IL 60077
708/676-3784

International Association of Business Communicators (IABC)
One Hallidie Plaza, Suite 600, San Francisco, CA 94102
415/433-3400

International Interactive Communications Society (IICS)
708/825-0816

Joint Ethics Committee
P.O. Box 179, Grand Central Station, New York, NY 10163
212/966-2492

Optical Publishing Association
P.O. Box 21268, Columbus, OH 43221
614/442-8805

Picture Agency Council of America (PACA)
1203 S. Maple Street, Northfield, MN 55057
800/457-7222

Picture Network International
2000 14th Street N., Arlington, VA 22201
703/558-7860

Printing Industries Institute (PII)
70 East Lake Street, Suite 300, Chicago, IL 60601
312/704-5000

Typographers International Association
84 Park Avenue, Flemington, NJ 08822
908/782-4635

COLOPHON

This book was typeset in Myriad, the first Multiple Master™ typeface family. "Multiple Masters" is a new form of PostScript typography, developed by Adobe Systems, Inc., for creating an unusually precise range of weights, widths, or other "axes of modification" supported by the particular type family. Myriad supports only weight and width, yet those tow alone can generate nearly a quarter of a million different fonts. The design challenge—as with most kinds of new publishing technology—is to use it in a restrained yet effective way.

AAA AAA **AAA AAA AAA**

Above are the 15 standard fonts that come with the package. These variations allow Myriad to retain its integrity in the great variations possible, unlike the distortions created with font manipulation as with an illustration program. Myriad's design is strongly reminiscent of Frutiger, a mono-stroked sans-serif European family designed in 1968. Myriad was designed at Adobe starting in 1990 by a team of type designers led by Robert Slimbach and Carol Twombly.

The body of the book was set in the regular-weight normal-width (400 RG 600 NO) of Myriad. The marginal items are in the same weight, but maximally condensed (400 RG 300 CN). For more information about Multiple Masters fonds, call Adobe Systems at 800/833-6687.

The cover and title pages were produced in Aldus FreeHand; the interior of the book was produced in Aldus PageMaker. All production was done on Apple Macintosh computers.

The book was designed by Michael Waitsman of Synthesis Concepts, Inc., Chicago.

ABOUT THE AUTHOR

Liane Sebastian is a workflow expert stemming from twenty years in the publishing industry. Specializing in strategic communications, Liane has been managed every aspect of electronic publishing. She served for four years as the President of ADEPT (Association for the Development of Electronic Publishing Technique), a national user group of desktop publishers. She was the President of Synthesis Concepts Inc., a graphic design firm specializing in corporate identity, publications, and collateral design for sixteen years, and now Synthesis Publishing. Liane is also currently evolving a complete graphics facility from conceptual strategy through prepress as the Division Manager and Creative Director for Smith, Bucklin & Associates, the world's largest association management firm. Liane is an award winning designer attaining national and international recognition. As an art director, marketer, project manager, writer, designer, illustrator, and desktop publisher, her experience spans every facet of graphic design. Additionally, Liane is a member of AIGA (American Institute of Graphic Artists), ACD (American Center for Design), The Graphic Artists Guild, CSAE (Chicago Society of Association Executives), and a founder of WIN (Women's Issues Network). She speaks at numerous conferences nationwide and helped to pioneer the CONCEPPTS Conference in Orlando. She has been featured in many articles in communications journals and magazines.